Environmental Public Health Policy for Asbestos in Schools

Unintended Consequences

Jacqueline Karnell Corn

Acquiring Editor: Arline Massey
Project Editor: Sara Rose Seltzer
Cover Design: Dawn Boyd

Library of Congress Cataloging-in-Publication Data

Corn, Jacqueline K.
Environmental public health policy for asbestos in schools : unintended consequences / Jacqueline Karnell Corn.
p. cm.
Includes bibliographical references and index.
ISBN 1-56670-488-X (alk. paper)
1.Asbestos -- Toxicology. 2. Asbestos abatement --Government policy--United States. 3. School bulidings--Safety regulations--United States. 4. Asbestos in building--Safety regulations--United States. 5. United States. Asbestos Hazard Emergency Response Act of 1986. 6. School buildings--Health aspects--United States. 7. Health risk assessment--United States--Case studies. I. Title
RA1231.A8 C67 1999
615.9'2539224--dc21

99-32020
CIP

This book contains information obtained from authentic and highly regarded sources. Reprinted material is quoted with permission, and sources are indicated. A wide variety of references are listed. Reasonable efforts have been made to publish reliable data and information, but the author and the publisher cannot assume responsibility for the validity of all materials or for the consequences of their use.

Neither this book nor any part may be reproduced or transmitted in any form or by any means, electronic or mechanical, including photocopying, microfilming, and recording, or by any information storage or retrieval system, without prior permission in writing from the publisher.

The consent of CRC Press LLC does not extend to copying for general distribution, for promotion, for creating new works, or for resale. Specific permission must be obtained in writing from CRC Press LLC for such copying.

Direct all inquiries to CRC Press LLC, 2000 Corporate Blvd., N.W., Boca Raton, Florida 33431.

Trademark Notice: Product or corporate names may be trademarks or registered trademarks, and are used only for identification and explanation, without intent to infringe.

© 2000 by CRC Press LLC
Lewis Publishers is an imprint of CRC Press LLC

No claim to original U.S. Government works
International Standard Book Number 1-56670-488-X
Library of Congress Card Number 99-32020
Printed in the United States of America 1 2 3 4 5 6 7 8 9 0
Printed on acid-free paper

FLORIDA GULF COAST
UNIVERSITY LIBRARY

Dedication

This book is dedicated to my grandchildren: Gabriel, Rachel, Aaron, and Zachary.

Foreword

Asbestos is one of many potentially toxic materials that under conditions of high exposure, caused pain and suffering in the working population. I watched a growing concern by the general population for the health consequences of low-level exposure to in-place asbestos in schools. How did our government, the scientific community, and the media communicate the well-known toxicological premise, "The dose makes the poison?" How would regulation deal with a lesser risk than that which took such a heavy toll of lives in asbestos workers? It was the unfolding of this drama that stimulated the writing of this book. Asbestos is, I believe, the first of many materials that have caused public concern based on the higher risks encountered in the workplace; lead is currently under scrutiny. I believe certain lessons could be learned from the asbestos experience.

The book was written with financial support provided by W.R. Grace Company to the Johns Hopkins University in the form of a research grant with me as the Principal Investigator. At the outset, Grace Company indicated that it did not want to see, review, or in any manner interfere with the progression of the research and writing of this book. They recognized this was a sensitive subject. I thank the W.R. Grace Company for adopting this policy. They were the least intrusive of any research sponsor that supported my research during my seventeen years as a faculty member at the Johns Hopkins University. Judgments, insight, and opinions in this book are mine and mine alone.

Individuals to whom I am indebted for their assistance and support are: Edward Terry, former Head Librarian at the Lillienfeld Library of the Johns Hopkins University School of Hygiene and Public Health; Cindy Rogers for her expert and devoted attention to the preparation of the manuscript; Dr. Michael Dellarco of the Environmental Protection Agency for extended discussions of risk assessment; and my husband, Dr. Morton Corn, for valuable counsel and insight into the scientific and regulatory establishments, and for his constant encouragement and support.

Table of Contents

Introduction

"History repeats itself — the first time as tragedy, the second time as farce."
(Karl Marx)

On January 3, 1977, Howell Township in New Jersey closed its elementary schools while scientists hired by The State Department of Environmental Protection tested to determine if dangerous levels of asbestos fibers existed in the classrooms and corridors of its schools. After reports of unexplained ailments, such as headaches, sore throats, and respiratory congestion in some school children, parents of 4,500 elementary school children threatened to boycott classes until the Board of Education removed asbestos-containing materials (ACM). In response to the parents' action, the Board of Education ordered schools closed at an emergency meeting held late in the night, and decided that children would attend school on a split session. Earlier, the Board of Education had decided to remove the asbestos during the summer vacation, but parents had expected removal even sooner, i.e., during the Christmas vacation.

Howard H. Clark, president of the Howell Township Board of Education, said he felt that the presence of ACM in the schools was a definite health hazard. Dozens of parents had called him to report their children's unexplained respiratory ailments. Mr. Clark stated that he would not send his own children to school until the township removed the asbestos.

The reported respiratory ailment of one Howell Township youngster led to all the activity. The boy's mother told officials her son suffered from headaches and a sore throat and that at times his neck glands grew bigger than his face. She said her other children were fine during summer months, but mysterious ailments began when they returned to school. She complained many times to the school officials, but she said, "Now it's time that some of the other parents come out and join me on this." Her son's pediatrician told her that the boy's ailment might have been caused by asbestos fibers flaking from the ceiling of his school.[1]

Four days later, New Jersey Governor Byrne stated at a press conference, "We must recognize the potential danger of sprayed-on asbestos." Joanne Finley, The Health Commissioner, Fred G. Burke, Education Commissioner, and David J. Bardin, Environmental Protection Commissioner, also attended the press conference. Burke told the press that 264 schools in New Jersey reported that some areas in their buildings contained asbestos.[2] Because they believed asbestos floating in air caused the Howell Township youngster's illness, concerned parents pressured the school board to close six schools while workmen replaced ceilings at a cost of 800,000 dollars. Cases similar to the one at Howell Township, with variations on the theme, were repeated throughout the 50 United States.

On February 19, 1977, a short article in the *New York Times* stated, "A thirteen year old student whose illness was attributed to asbestos ceilings and led to school closings and a statewide search for cancer-causing fibers has been sick with mononucleosis his doctor said today."[3]

Why did parents respond to asbestos in schools in such an irrational manner? Was fear of ACM in the schools exaggerated? Did scientific evidence exist that indicated asbestos in schools caused school children to become ill? What was the effect on public policy? Was all relevant information including the impact of decisions considered? What would be the aftermath of the Howell Township and other school systems response to the asbestos in their buildings? The purpose of this book is to address these not so easy-to-answer questions, by focusing on the development, institutionalization, and consequences of federal environmental policy for asbestos in schools.

Essential scientific and technical questions which should have guided policy decisions were often overlooked and political considerations prevailed. In this book I have endeavored to focus both on political and technical issues to present as clear a picture as possible of how asbestos policy for schools was formulated. A key question at the heart of the controversy was, Does ACM in a building increase heath risks to the building occupants? To answer that question the Environmental Protection Agency (EPA), the lead federal agency dealing with asbestos in schools, departed from utilization of the occupational health/industrial hygiene approach to defining whether a potentially toxic material, i.e., asbestos, in place in a building can affect building occupants. The well-known paradigm or model in the occupational and environmental health field developed by industrial hygienists includes recognition, evaluation, and control of potentially hazardous material. That concept for ACM in schools, although overlooked in initial policy formulation, has helped to inform me in my quest for an answer as to how and why public policy for asbestos in school buildings was developed. In this paradigm, recognition refers to the presence of the material; evaluation involves measuring the amount inhaled, swallowed, or absorbed through the skin and whether or not these amounts are damaging to health; and control refers to measures to reduce the amount of material present, if evaluation indicates control is necessary. Agency departure from this paradigm led to public misunderstanding, confusion (sometimes hysteria), needless removal of ACM from school buildings, expenditures of huge sums of money, and even conflict among scientists over values, as well as science. In other words, it led to unintended consequences.

Currently, United States government policy favors keeping management of ACM in place. The effort to steer policy back to the proven established paradigm or model of the occupational health and industrial hygiene professions is part of this story. It began with the measurement of the asbestos in the air of school buildings to determine how much occupants breathe, in order to estimate the level of risk, if any, to building occupants. But it took almost two decades to both understand and disseminate knowledge related to health effects of ACM in buildings. Today we know the risks are minimal and that removal of ACM usually is the least desirable option for dealing with asbestos in buildings.

Asbestos is a mineral with a long history. Both the uses and understanding of the benefits and risks associated with it have dramatically changed over time. Despite its long history, there had never been as much public controversy about asbestos as that generated by the politically-charged and difficult to resolve issue of asbestos in schools. Scientists, members of trade associations, trade unions, environmentalists, industrialists, the courts, realtors, educators, and the general public were caught up in fierce debates which reflected competing interests, generated public fear, and polarized all concerned. What appeared on the surface to be a scientific and methodological dispute manifested itself as a political and philosophical conflict.

Nevertheless, very little historical literature has been dedicated to the politics of non-occupational asbestos, although the scientific and technical aspects of asbestos have received more world wide attention than has any other toxic agent. More is known about asbestos toxicity and the dose-response curve than any other industrial material.[5] Most of what we know about adverse health effects associated with asbestos utilization derives from past high levels of workplace exposures. In the 1970s, concerns about health hazards related to asbestos spread from occupationally exposed populations to those experiencing non-occupational exposure. By the end of the 1970s, school children and teachers were targets of concern, and beginning in 1979, EPA assisted school administrators in identifying and controlling ACM in their buildings.

In 1971, EPA labeled asbestos a hazardous air pollutant as defined by the Clean Air Act and issued regulations to control asbestos emissions from manufacturing, milling, roadway surfacing, and demolition projects. In 1973, EPA banned the spraying of ACM in buildings under the National Emission Standards for Hazardous Air Pollutants (NESHAPS). Almost ten years later (1982), under the Toxic Substances Control Act authority, the agency issued the Asbestos-In-Schools Identification and Notification Rule. It required inspection for friable asbestos in all schools and notification to parents, teachers, and schools workers. Congress passed the Asbestos School Hazard Abatement Act (ASHAA) in 1982 to provide financial assistance to needy schools with serious asbestos hazards. On October 22, 1986, President Ronald Reagan authorized the Asbestos Hazard Emergency Response Act (AHERA); it required EPA to "provide for the establishment of federal regulations which require implementation for appropriate response actions with respect to asbestos-containing material (ACM) in the nation's schools in a safe and complete manner." By law, EPA was to "define the appropriate response action" and describe "the least burdensome methods which protect human health and the environment." All public and private schools would be required to inspect for asbestos, develop asbestos management plans, submit them to the state, and implement appropriate response actions and abatement.[6]

Representative James Florio of New Jersey won unanimous passage of AHERA, which he proposed and sheparded through Congress. AHERA represents a policy decision with far-reaching social, economic, educational, political, and scientific repercussions. Its history raises questions of causation, acceptable risk, seriousness of the hazard, veracity of the scientists associated with asbestos research, responsible regulatory options, as well as questions about the roles of Congress, environmentalists, labor organizations, scientists, the press, EPA, the public, and even the courts in creating environmental policy. It questions the uses of information and the impact of public information presented, withheld, or distorted in the making of environmental policy.

By 1988 an EPA report estimated that approximately 20% of public and commercial buildings in the United States with ACM undergo some form of abatement activity each year, and about 7% have all ACM removed. At this rate, according to the study, asbestos could be abated from American schools in 10 to 15 years. The EPA also estimated that 733,000 public and commercial buildings contained friable ACM, and that between 160,000 and 255,000 abatement jobs are performed each year on buildings of all types. They also found abatement activity in schools much higher than in public buildings (28% of schools that had asbestos abatement had full removal). The report stated that 65% of abatement activity consisted of removal, 13% involved repair, 12% encapsulation, and 10% enclosure, and that most abatement occurred in prosperous regions.[7]

Billions of dollars were spent to remove asbestos from buildings and a major industry was built around that effort. Although removal continued to take place, asbestos policy was increasingly questioned. In 1990, the once-applauded asbestos removal activity was called, "the asbestos removal fiasco." Those who questioned the policy of asbestos removal suggested that billions of dollars had been spent needlessly and billions more would be wasted. The economic and social costs of AHERA have been enormous. Scarce resources needed for better education were allocated for asbestos abatement. It also greatly affected the scientific establishment. Eminent authorities argued on both sides of the issue, eroding the idea of impartial and objective science. Both proponents and opponents of removal from school buildings showed they were capable of distorting science to achieve their own aims. Asbestos tarnished the image of the disinterested and objective investigator (whether accurate or not) who seeks knowledge to apply it for the public good. Even more disturbing to contemplate is the fact that many of the workers hired to remove asbestos worked in unsafe and unhealthy conditions, often unaware that they were exposed to unacceptable concentrations of asbestos fibers.

Environmental issues in general, and asbestos in particular, reveal disagreements, inconsistencies, and uncertainties in the scientific community. In the case of asbestos, some scientists did not agree on health effects, or on assessment and measurement methods. Different scientists analyzed the same facts and reached different conclusions. As a result, scientific controversy was transformed into a vituperative public debate which extended beyond causal matters to reflect values and politics.

When concerns were occupational, asbestos did not excite too many people. But later, non-occupational exposure to asbestos aroused public indignation because it came closer to home, had an immediate connection to personal life, and perhaps more important, involved school children. One of the major concerns of this book is to explain how the issue of non-occupational asbestos came to be viewed as a public health problem, including how public health and environmental priorities were set and how decisions made and executed.

The working hypothesis of this book is that when the empirical base of science utilized for regulatory purposes is uncertain, controversial, or in question, the manner in which the public and decision makers respond to the issue becomes dependent on other factors especially political, beside science. The presentation and perception of the issues then assumes great importance in determining response. In the case of determining a policy for asbestos in schools and other public buildings, the legacy of distrust resulting from past actions (asbestos in the workplace), combined with scientific uncertainties, conflicting values and views, polarization of scientists and the public on environmental issues, fear, and the regulatory climate of the 1980s led to exaggerated, poorly-focused, public and governmental (including the agency charged with regulating asbestos) concern, and misunderstanding, which led to a policy for asbestos in schools and other public buildings that was and still remains costly and controversial.

In order to achieve a defensible and less controversial regulatory effort, scientific and technical information, essential to public understanding of environmental issues, needed to be provided early and in clear terms to the public, Congress, and potential regulators and regulatees. In the case of asbestos in schools, that information did not become available. Bitter debates ensued, costs soared, and the courts intervened.

Asbestos, a toxic agent initially encountered in the workplace, caused workers to suffer major adverse health effects. It was later measured in the community at concentration levels of magnitude less than those to which workers had been exposed. Lack of understanding of toxicological principles, in this case the most important of which is dose-response, and inadequate or no risk assessments, led to exaggerated public concern and misunderstanding of the issue. Vigorous educational efforts did not precede public awareness of the issue, and the scientific community reluctantly engaged in the process of public communication and clarification of issues. Public demand for remedial action, based on inadequate information and misunderstanding, coupled with fear and a crisis mentality, precipitated premature, poorly-conceived, and imprudent public health policy. Such a policy left the public vulnerable to opportunistic commercial entrepreneurs who benefitted from profitable remedial activities that were unnecessary at best, and increased health risk to the public at worst.

My own view as a historian is that understanding and insight into past policy including how, why, and who caused action to be taken, can enlighten and even guide the scientific and regulatory communities when faced with a future need to respond to other low level environmental contaminants. Thus, I asked the questions: "How did this happen?," "Why did it happen?," and "Can it happen again?" The answers are as complicated as the questions are simple. The story of asbestos is a cautionary tale. Other toxic agents, i.e., lead contamination of homes, and indoor pollutants such as nitrogen dioxide or radon, could follow the model of asbestos, raising similar questions.

The format of the book is both topical and chronological. It is divided into three sections: "Background and Perspectives," "On the Road to AHERA," and "Unintended Consequences." "Background and Perspectives" includes an introductory chapter and background material necessary to understand the issues. Chapter 1, "A Profile of Asbestos, the Magic Mineral," describes properties and uses of asbestos and presents a historical perspective on asbestos and disease in the workplace. Chapter 2, "From Occupational to non-occupational Exposures," deals with the changing emphasis on asbestos exposures. It includes a discussion of risk assessment and its role in determining environmental policy. "On the Way to AHERA" describes and analyzes how and why the AHERA was passed. Chapter 3, "The Making of An Environmental Issue," discusses how asbestos in schools became an issue. Chapter

4, "Before AHERA" and Chapter 5, "Congress Passes AHERA," explore how AHERA was passed and the proposed rulings. "Environmental Policy Run Amuck" is about the aftermath of AHERA. It points out issues, inconsistencies and problems created by AHERA and EPA, the agency given the responsibility for enforcing the act. Chapter 6, "After AHERA," is meant to point out the issues, inconsistencies, and costliness of the law. Chapter 7, "Dueling Conferences," discusses two "scientific conferences" with different viewpoints, once again pointing out issues and inconsistencies. Chapter 8, "Summing Up," addresses the hypothesis presented earlier in the introduction and presents the author's conclusions. In all chapters, I have endeavored to depict the role played by various actors in this drama: scientists, school administrators, labor unions, the Congress, the courts, environmentalists, regulatory agencies, and ordinary citizens in the resolution of a public health and environmental issue, asbestos in schools.

The literature on risk analysis and risk perception along with the scientific literature helped to inform and guide this inquiry. For example, *Risk Assessment in the Federal Government: Managing the Process*[10] reflected the movement in federal agencies in the early 1980s toward more clearly defining health risks and acceptance of risk assessment as an analytical strategy for regulatory decision making. The document also noted the need to differentiate between risk assessment and risk management and to develop methodology. In the 1970s, W.W. Lowrance, author of *Of Acceptable Risk*,[11] indicated that decisions were made in both a social and scientific framework noting that addressing risk involves two extremely different kinds of activity: (1) assessing risk, an empirical, scientific activity, and (2) judging risk, a normative political activity. The first implies objectivity, while the second implies social value.

I found Kelman's interpretation in *Making Public Policy*[12] of the policy-making process in the United States, a helpful and optimistic account of the political process of decision making in major institutions of the federal government, including the Congress, the Presidency, the Bureaucracy, and the Supreme Court. The research of both Nelkin and Jasonoff has also provided a number of insights for me to build on in the area of the social dimension of risk and the legal and political analysis of risk, including the exploration of how political alignments underlay scientific controversies. Nelkin indicated that the management and assessment of risks involves perception and priorities that are shaped by economic, political, and ideological concerns and that defining risk management is fundamentally controversial.[13] Jasonoff explored broad issues of science policy and politics showing how particular legal and institutional features influence the interpretation of scientific data and the resolution of technical controversies.[14] I have also found Lave's work in risk analysis and the utilization of quantitative information for decision making useful.[15]

The subject of perception of risk is a broad one and the current literature which utilizes sociological, psychological, and political science concepts has been helpful. A number of researchers have stressed this approach to answer questions about the dynamics of risk perception and selection. Their conclusions and insights also informed this research. For example, Douglas and Wildavsky[16] have shown that the risks people decide to focus on are not chosen simply to protect their environment or health. They make choices based on their beliefs about values, social institutions, and moral behavior. Furthermore, people exaggerate or minimize risks according to the social, cultural, and acceptability of the underlying activities. The work of others, who utilize the psychological approach, indicate that lay people often have difficulty understanding probabilistic information, especially when the probabilities are small and the risk unfamiliar, and that people often overestimate the risks of a dramatic cause of death. They have indicated the need to integrate psychological with social and cultural factors, and that concerns about risk are largely based on qualitative factors rather than quantitative.[17]

I am indebted to other historians who have pursued environmental history, especially Thomas Dunlap who utilized historical method to present a history of DDT and the making of public policy for that infamous insecticide,[18] and Samuel Hays for his insight into environmental politics.[19] They add the dimension of historical context to environmental studies, one that is too often overlooked.

The clear message, then, is that in order to understand the different perceptions of environmental risk and societal response to risks, one must be cognizant of values and of the social, political, and historical context, as well as the scientific context in which they are developed.

REFERENCES

1. Narvaez, A.A., Eight schools in Jersey are closed to check asbestos fiber levels, *New York Times,* p. 31, Jan. 4, 1977.
2. Narvaez, A.A., Byrne discounts asbestos peril, *New York Times,* p. 9, January 8, 1977.
3. Illness wasn't due to asbestos, *New York Times,* p. 51, Feb. 19, 1977.
4. Corn, M., Asbestos and disease: an industrial hygienists' perspective, *Amer. Ind. Hyg. J.,* 47(9), 515, 1986.
5. *Federal Register,* 47 FR 23363, May 27, 1982, *Friable Asbestos-Containing Minerals in Schools; Identification and Notification.*
6. *Federal Register,* October 30 (52 FR 41826) Final Rule, Oct. 30, 1987.
7. USEPA, EPA Study of Asbestos Containing Materials in Public Buildings: A Report to Congress, *U.S. EPA,* Washington, D.C., (560/5-88-0222), 1988.
8. The asbestos removal fiasco, *Science,* 1990.
9. Science and faddism, *Wall Street Journal,* 1990.
10. National Academy Press, National Research Council, *Risk assessment in the federal government: managing the process,* Washington, D.C. 1983.
11. Lowrance, W.W., *Of Acceptable Risk: Science and the Determination of Safety,* William Kaufman, Inc., Los Altos, CA, 1976.
12. Kelman, S., *Making Public Policy,* Basic Books, Inc., New York, 1987.
13. Nelkin, D., *The Language of Risk,* Sage Publications, Beverly Hills, CA, 1985.
14. Jasonoff, S., *Risk Management and Political Culture,* Russell Sage Foundation, New York, 1986.
15. Lave, L.B., *The Strategy of Social Regulation: Decision Frameworks for Policy,* Studies in the Regulation of Economic Activity, The Brookings Institution, Washington, D.C., 1981.

16. Douglas, M. and Wildavsky, A., *Risk and Culture: An Essay on the Selection of Technologies and Environmental Dangers*, University of California Press, Berkeley, CA, 1982.
17. Fischoff, B., Lichtenstein, S., Slovic, P., Derby, S.L., and Keeney, R.L., *Acceptable Risk*, Cambridge University Press, New York, 1981.
18. Dunlap, T.R., *DDT-Scientists, Citizens, and Public Policy*, Princeton University Press, Princeton, NJ, 1981.
19. Hays, S., Beauty, *Health and Permanence — Environmental Politics in the United States, 1955–1985*, Cambridge University Press, New York, 1987.

Background and Perspectives

1 A Profile of Asbestos, the "Magic Mineral"

HISTORICAL PERSPECTIVE

Awareness of asbestos' unique properties, the number and variety of its applications, and its health effects must inform any discussion of the history of asbestos and its associated policies, including past and present protective measures. Asbestos exhibits contradictory qualities. The mineral has done much good, and at the same time, caused great harm. The paradoxical nature of the "magic mineral" manifested itself almost simultaneously with early utilization.

Asbestos represents a class of controversial environmental problems in our modern industrial society at the interface between science and politics, one that requires both scientific and political understanding. It illustrates the dilemma an industrialized society faces when deciding whether or not to continue utilization of a valuable, but nevertheless potentially toxic, material. Along with this dilemma, scientific uncertainties create a set of complex and unresolved social, scientific, and political issues: (1) determining the extent of asbestos related disease; (2) controlling the hazards; (3) regulating the material; (4) controversy over causation of disease; and (5) establishing criteria on which to base a standard.

In the 1980s, litigation based on personal injury as a consequence of asbestos exposure in the workplace burgeoned. The litigation illustrated the controversial and contentious nature of asbestos issues. By 1984, 25,000 workers or their survivors had filed lawsuits against companies in the asbestos industry. The lawsuits included cases claiming federal government liability for alleged negligence in establishing adequate workplace standards. Before 1983, asbestos workers filed most personal injury lawsuits (i.e., electricians, mechanics, shipyard workers, and miners). However, in 1983, a new group of plaintiffs appeared who had worked on buildings insulated with asbestos. In the same year, school districts sued asbestos manufacturers and suppliers for the cost of removing and replacing school materials that contained asbestos.

The early image of asbestos as a much-needed, beneficial material had already begun to erode in the 1960s. By the 1970s its use as a building material was definitely in question. The risk of asbestos, present from the start, but becoming better known, caused some people to question its intense utilization in the building trades.

Donald Hunter, author of *The Diseases of Occupations,*[1] an occupational health textbook first published in 1955, discussed the ancient history of asbestos tracing it back at least 2,000 years.

> On a small scale asbestos was made into cloth at least 2,000 years ago. Herodotus (450 BC) relates how the Romans mined it in the Italian Alps and Ural Mountains, using it for enshrouding corpses before cremation to permit easy collection of the ashes for burial. They named it amianthus, meaning without miasma, undefiled or incorruptible. Pliny (50 AD) refers not only to difficulties in weaving asbestos, but also to the use of respirators to avoid inhalation of dust. Strabo (30 BC) and Plutarch (AD) refer to the wicks of the lamps of the Vestal Virgins as asbesta, the unquenchable, inextinguishable or inconsumable. It is stated that Charlemagne possessed a tablecloth made of it, which was cleansed by passage through fire.[2]

This mysterious fire-resistant quality probably gave asbestos the name, "magic mineral." Other writers have said that the ancient Greeks, Romans, and Egyptians knew about the fire-resistant qualities as well as the hazardous nature of asbestos. The "magic mineral" remained merely a curiosity until the 19th century when developing technology necessitated insulation of hot engines, boilers, and piping, and created many new industrial applications. At the time, the benefits of asbestos usage, when contrasted to the risk incurred, outweighed the known hazards.

Besides resistance to heat, asbestos has many other desirable qualities. Tensile strength, durability, and resistance to wear, corrosion, and chemicals all add to its utility. Asbestos can be pressed to form paper or structural reinforcement materials such as cement, asphalt, or plastic. It can be spun into yarn, woven into fabric, and braided into rope. It contains attributes so desirable and beneficial that between 1900 and 1980 some 36 million metric tons were used in over 3,000 products.[3] Much of it was introduced without significant precaution. Asbestos was, after all, inexpensive, widely available, easily fabricated, strong, fire-resistant, and chemically stable.

At early stages of technological growth, involving the use of a needed hazardous material the chance of jeopardy to health incurred by utilization is often accepted. In some cases the perils are unknown. As technology becomes more refined, and those who utilize the material become more aware of the hazards, and as social philosophy changes, attitudes toward risk also becomes more humane. New awareness of danger and the concomitant change in social philosophy can cause a reevaluation of the continued utilization of the material.[4] This occurred after hundreds of years of asbestos usage, when asbestos became the subject of intense public interest in the 1970s.

USES OF ASBESTOS

The modern asbestos industry, which developed between 1912 and 1982, includes the following products: floor tiles, gaskets, packing, friction products, paint coatings and sealants, asbestos-reinforced plastics, asbestos-cement pipe, paper, and cement sheet. After 1982, asbestos consumption rapidly declined. However, asbestos had also been added to a wide variety of construction materials which made it present in many school buildings, commercial and public buildings, and homes.[5] The use of asbestos in building materials started to decline in 1973 when EPA banned certain applications, but the large amounts of asbestos already in place remained a possible source of exposure throughout the lifetime of millions of buildings. Asbestos in place could be regarded as a potentially hazardous material because it had the possibility to become airborne and enter the lungs.

Large-scale asbestos mining and commercial production started early in the 20th century and greatly accelerated during World War II. Asbestos consumption in the United States rose from less than 100,000 tons in 1912 to approximately 750,000 tons during World War II, and 800,000 tons in the 1970s.[6] During World War II, when the most rapid increase in asbestos consumption occurred, shipbuilding and ship repair consumed huge amounts of asbestos. At the height of asbestos utilization, industries in the U.S. that manufactured asbestos products or used asbestos-containing products employed over 37,000 men and women in the manufacture of primary asbestos products (i.e., production of manufactured goods from raw asbestos fibers), and 300,000 people in secondary asbestos industries (i.e., processing asbestos-manufactured products to make other products). Millions more people worked in asbestos consumer industries, i.e., utilized a finished product containing asbestos. For example, 185,000 people worked in shipyards and almost 2,000,000 worked in automotive sales, service, and repair.[7] It has been noted that the widespread use of asbestos as an insulator in buildings and in break linings may have saved hundreds of thousands of lives.

From the end of World War II up to the 1980s, all states specified that asbestos be used as a building insulation material. From 1930 to the mid-1970s, builders extensively applied asbestos, mostly amosite and chrysotile, individually and in blends to building structures, walls, ceilings, and on thermal systems within buildings to provide insulation and fire-proofing. In fact, most of it was prescribed by building codes.[8] Sprayed-on asbestos protected structural steel beams from the potential melt down effects of high temperatures. It also permitted the egress of occupants and provided ceilings and interior surfaces with insulation to retard fire.

Occupational exposures to asbestos fibers occurs during mining, milling, production, processing of asbestos-manufactured products from raw fiber (primary industries), processing of asbestos-manufactured products to make other products (secondary industries), and utilization of finished products containing asbestos without further modification (consumer).

ASBESTOS DESCRIBED

The generic term asbestos refers to a group of naturally occurring, flexible, fibrous, hydrous silicate minerals divided, on the basis of mineralogical features, into two groups: serpentine and amphiboles. Asbestos describes a property known in mineralogy as crystal habit, the fibrous aspect or asbestiform aspect in which some minerals crystallize. A number of minerals can crystallize in an asbestos form, aspect or habit, but only six have industrial uses: chrysotile, crocidolite, amosite, anthorphyllite, tremolite, and actinolite.[9] Only chrysotile, the most abundant, belongs to the serpentine group. Crocidolite and amosite, the other two most commonly used asbestiform minerals, along with anthorophyllite, tremolite, and actinolite belong to the amphibole group.

Chrysotile crystallizes as parallel fibers that can be separated easily into individual fibers. It is composed of pliable curly fibers and differs structurally and chemically from amphiboles. The amphibole group of minerals is composed of a number of structurally similar, but chemically different, minerals. Unlike the curly chrysotile fibers, amphibole asbestos fibers are straight and needle-like.

Chrysotile asbestos has a spirally-wound or concentric-layered silicate structure. The more brittle amphibole fibers split longitudinally more readily than does chrysotile, are dustier, and have a greater tendency to become airborne.[10] These characteristics contain the key to an understanding of one of the controversial scientific questions about asbestos: Is chrysotile asbestos a less potent carcinogen than the amphiboles amosite and crocidolite? The relationship between the carcinogenic potential of fibers and their fibrogenic properties is not yet fully understood and scientific evidence that relates the type of asbestos inhaled to the manifestation of disease remains highly controversial.[11] In the United States, the occupational permissible exposure limit for asbestos does not differentiate among types of asbestos, while other countries differentiate among fiber types and invoke different permissible exposure limits (PEL) for different fiber types.

Chrysotile dominates commercial asbestos consumption. 1980 production figures indicate that it accounted for approximately 90% of world production of asbestos. Canada and the former Soviet Union contain major chrysotile deposits. Other countries that mine chrysotile asbestos include: South Africa, Zimbabwe, Cyprus, Italy, The United States, China, Switzerland, and Brazil. Commercial amphibole asbestos comes almost solely from South Africa; crocidolite is mined in the Cape and Transvaal regions of South Africa and in Bolivia. In 1980, amphibole asbestos accounted for 6% of the world's asbestos production. As of 1980, amosite mined in the Transvaal and India accounted for approximately 3% of world production.[12] Apart from commercial deposits asbestos can be found everywhere in the earth's crust and its surface waters. The mineral is ubiquitous.

ASBESTOS AND DISEASE

The paradox of this ubiquitous "magic mineral" is that while it is so utilitarian, it can also cause serious, debilitating illness and death. The major pathological effects of asbestos result from inhalation of asbestos fibers suspended in ambient air. Most of our knowledge about asbestos and disease has been derived from the workplace. Medical literature began documenting disease resulting from inhalation of asbestos dust in the workplace beginning in the early 20th century.[13] The major diseases associated with asbestos inhalation include: asbestosis, lung cancer, and mesothelioma.

Asbestosis, a chronic, restrictive lung disease caused by inhalation of asbestos fibers was the first known disease related to asbestos exposure. It is associated with heavy occupational exposure to asbestos. Mesothelioma, a rare cancer of the surface lining of the pleura (lung) or peritoneum (abdomen) generally spreads rapidly over large surfaces of either the thoracic or abdominal cavities. No effective treatment exists for mesothelioma. It occurs among insulators, those who work in asbestos plants, and shipyard workers, and has been reported among persons living in the same house as asbestos workers and in the neighborhood of asbestos mining and milling.[14] Similar to asbestosis, mesothelioma is specifically linked to asbestos. Unlike asbestosis and mesothelioma, lung cancer is not specifically associated with asbestos. Lung cancer also has a history of association with cigarette smoking and atmospheric pollution. Although it is recognized that asbestos, in the absence of cigarette smoking, can induce lung cancer, issues of causation are often raised when lung cancer develops in asbestos workers who smoke.

Asbestos disease manifestations have been linked to the amount of asbestos affected persons inhaled. Generally, larger quantities of inhaled asbestos or high dose has been related to asbestosis, while lower quantities or low dose is linked to mesothelioma and lung cancer. Currently there is controversy over the relationship of certain types of asbestos to mesothelioma.[15] It is generally accepted that the type of mineral, dimension of the fibers, concentration of fibers, and duration of exposure can influence the occurrence of asbestos disease.

RECOGNITION OF PATHOGENICITY

Recognition of asbestos pathogenicity came slowly. The continuously altering perception of the risks to health demonstrated both developing new knowledge about asbestos disease and changing societal attitudes toward asbestos risk. Approaches to assessing inhalation risk from asbestos underwent profound change as well.[16] One of the most influential factors involved in the growing understanding of asbestos pathogenicity was increased utilization of the mineral. Other factors include: new technologies, new measuring instruments, new biological data, regulatory imperatives, and redefinition of health risks associated with asbestos.

Even before the 20th century, early warnings about the health hazards associated with asbestos appeared. However, usage was, relatively speaking, minimal, the known number of workers affected small, and occupational disease not generally a matter of social concern. The benefits of utilizing asbestos seemed to outweigh the known hazards. Today it is a well-known and universally accepted fact that asbestos exposure can cause serious illness and death and that the major pathologic effects of asbestos result from inhalation of fibers suspended in air. Documentation, including physicians' reports, of the association between asbestos and disease in the workplace began early in the 20th century. By 1965, more than 700 articles had appeared in worldwide literature detailing hazards of asbestos exposure.[17]

Asbestos textile workers suffered from the first reported cases of asbestos disease. In France in 1906, a factory inspector named Auirbault recorded 50 deaths that occurred during a 5-year period (1890–1895) among workers at an asbestos weaving mill.[18] Dr. H. Montague Murray reported the death of a 33-year old woman from fibrosis of the lung due to inhalation of asbestos dust to a parliamentary committee on industrial disease compensation in 1907.[19] Nevertheless, scant knowledge about asbestos and its associated risks, coupled with the small amounts of the mineral used, limited observations and understanding of the relation between asbestos and disease. In fact, the fibrotic disease described by Murray, Auribault, and others did not even receive a name until 1927 when Dr. W.E. Cooke named it asbestosis.[20] Earlier in 1924, Cooke had described the death from pulmonary fibrosis of a woman who worked in a textile mill for 20 years. The first account of asbestos-related disease in the U.S. appeared in medical literature in 1917.[21] By then, physicians in industrialized nations knew of the disease, asbestosis, as a specific chronic disease of the lungs attributed to breathing asbestos. In medical journals they described cases of asbestosis including symptoms of the disease, its natural causes, and the pathological changes visible at autopsy. As the production of asbestos began to expand, the number of observations of asbestosis increased. In 1928 and 1929 the British government investigated the condition of textile factory workers. They reported in 1930 to the Parliament that, "... inhalation of asbestos dust over a period of years results in the development of a serious type of fibrosis of the lungs."[22] They recommended dust suppression. British asbestos industry regulations followed in 1931.[23]

Clinical reports in the U.S. confirmed occurrence of asbestosis among asbestos workers. Frederick Hoffman, chief actuary for the Prudential Life Insurance Company, wrote *Mortality from Respiratory Diseases in the Dusty Trades* for the United States Bureau of Labor Statistics.[24] Hoffman acknowledged that levels of dust in industry caused a considerable dust hazard, and American and Canadian life insurance companies generally declined to insure asbestos workers based on their assumption that injurious health conditions existed in that industry. Clinical reports continued to confirm cases of asbestosis among asbestos workers. They appeared in *The Journal of the American Medical Association,*[25] *The American Journal of Public Health,*[26] and *The Journal of Industrial Hygiene.*[27] In 1935, *Public Health Reports*[28] published the results of a Metropolitan Life Insurance Study and in 1938 *Asbestosis in the Asbestos Textile Industry,* often referred to as the Dreessen study, was published.[29] Asbestos workers made the first formal claims in the U.S. for compensation for asbestos disease in 1927. The Johns Maniville Company settled all claims against it out of court in 1933. In the 1930s, public health and medical publications had begun to incorporate the topic of industrial hazards and included asbestos dust among those hazards.[30]

By the 1940s, physicians and industrial hygienists had identified asbestos dust as dangerous and unhealthy and acknowledged that inhalation of that dust caused asbestosis, a fibrotic lung disease. Twenty years elapsed between the initial report of asbestos as a cause of fibrotic disease and the general acceptance of asbestos dust as a health hazard. Even though the scientific literature on asbestos hazards continued to accumulate, workers exposed to the fiber received little or no protection. Knowledge had not yet been translated into effective action. Asbestosis is associated with the high levels of exposure in the workplace which occurred well into the 20th century when lack of both engineering controls and interest in worker health, along with increased utilization of the mineral, exposed workers to heavy concentrations of asbestos dust. American industries used more and more asbestos, devising new applications, and seldom considering the health of asbestos workers.

The first reported case of the association between asbestos dust and lung cancer appeared in 1935,[31] but the first rigorous epidemiological study appeared in 1955.[32]

To complicate the matter, the latency period for cancer lasted longer than for fatal asbestosis and lower levels of dust could cause cancer. The foreshadowing of the tragedy to come had already begun in the 1930s when British, American, German, and French reports associated asbestos dust exposure with the development of lung cancer. In 1935, Lynch and Smith published a report of carcinoma of the lung in a 57-year old man who worked for 27 years as an asbestos mill weaver in an extremely dusty atmosphere.[33] In Great Britain during the same year, Gloyne reported two cases, both women; each had asbestosis and small carcinomas.[34] The British *Annual Report of the Chief Inspector of Factories for the year 1947* reviewed all reported cases of workers who died as a result of asbestosis since the first reported cases in 1924. Of these workers, 13% had lung cancer, but the expected incidence was 1%.[35] Doll's classic epidemiological study documenting the high risk of cancer to asbestos workers only came about 8 years later. Finally, after 3 decades, the 1964 landmark Conference on Biological Effects of Asbestos convened by the New York Academy of Sciences resulted in a consensus among scientific investigators that asbestos could cause lung cancer.[36]

The lengthy route from initial cognizance to confirmation, to general acceptance of the association between asbestos and lung cancer, similar to that between asbestos and asbestosis, took decades. Tragically, large numbers of workers would, in the meantime, become desperately ill and many of them would die from a painful disease.[37] Furthermore, in the 1960s it became increasingly clear that risk of disease was not confined solely to workers in mining and textile manufacturing. It extended to those who used asbestos products: shipyard workers, insulation workers, and many others outside fixed location industries.

As early as the 1940s and 50s, scattered medical reports of a third and rare disease, mesothelioma, caused by asbestos exposure, appeared. In the 1960s, Wagner and associates clearly established the association between asbestos and mesothelioma.[38] It has since been confirmed and demonstrated in the laboratory, epidemiological studies, and case reports.

Thus, by the 1960s the hazards of asbestos in the workplace were recognized by a segment of the medical profession and in technical and scientific circles. Although knowledge about health effects accumulated slowly, it was never secret. A body of scientific and medical literature about diseases associated with asbestos, techniques for measuring dust, and control technology existed. An international literature existed that included journals, government publications, and textbooks. The academic, occupational health, and public health communities knew about the relationship between occupational disease and asbestos dust. Early in the 20th century a knowledge base had been developing. It started with medical case studies and later utilized epidemiological, toxicological, and medical information that accumulated together with more and more sophisticated measurement and control techniques. All this knowledge about asbestos health risks was learned from the workplace. But American society, as distinct from a relatively small core of professionals, occupational physicians, industrial hygienists, public health, and technical professionals knew little about asbestos and the health risks associated with asbestos inhalation. They were poorly informed and certainly not sensitized to the idea that a toxic substance in the workplace could contribute to future illness. It is no wonder then, that with so much knowledge at hand, so little was done prior to the 1970s to control asbestos in the workplace.

RESPONSE TO ASBESTOS IN THE WORKPLACE: TLVS

Why hadn't the problems associated with asbestos entered the public debate? The asbestos industry has been faulted for not allowing access to its plants, influencing the research agenda, controlling publication of results of scientific studies, failing to provide warnings to users of products, and entangling reputable scientists in a web of deceit and manipulation. Industrial hygienists and occupational physicians (the professionals) have also been blamed for inattention to asbestos hazards as well as to being preempted by the asbestos industry.[39] Perhaps the fact that no federal authority existed to regulate the workpace can also answer the question. Professionals before 1970 depended on persuasion rather than government authority to make a workplace safe. Lack of social controls made it possible for asbestos companies to meet their own goals at the expense of workers' health and industry controlled the workplaces where neither health professionals or the federal government counter-balanced the power of industry. The absence of federal regulation of the occupational environment until the Occupational Safety and Health Act was promulgated in 1970 contributed to the inertia in recognizing asbestos as a causative disease agent for cancer.

Prior to the Occupational Safety and Health Act of 1970 there were suggested professional guidelines for asbestos known as Threshold Limit Values (TLVs).[40] In 1938, Dreessen and others studied the North Carolina asbestos textile industry. The United States Public Health Service published *Asbestos in the Asbestos Textile Industry,* featuring results of the study. The study's authors recommended dust control after they observed the process that produced dust in the asbestos textile factories. They concluded that: (1) the percentage of people in different occupational groups affected by asbestosis or its symptoms varies with average dust concentrations and length of employment; (2) only three doubtful cases of asbestosis were found below five million particles per cubic foot of air, and they were diagnosed as doubtful; (3) well-established cases occurred at higher concentrations; and (4) if asbestos dust concentrations in the air breathed are kept below this limit new cases of asbestosis will not appear. They concluded that because clear cut cases of asbestosis were found only from dust concentrations exceeding 5 million particles per cubic foot of air (mppcf) and none were found at lower concentrations, 5 mppcf could be regarded "tentatively" as the threshold value for dust exposure until later data are available.[41]

Today, the Dreessen study is considered a flawed and limited epidemiological investigation. However, the number for a "safe" level of asbestos in air stood at 5 mppcf for 30 years as the TLV guideline. The guideline was not seriously reappraised until after the 1964 meeting of the New York Academy of Science and publication of its proceedings, *Biological Effects of Asbestos*, which confirmed asbestos as a carcinogen.

The critical assumption of the TLV concept is that a level of exposure exists for noncarcinogenic substances below which no adverse health effects will occur. In contrast, U.S. regulatory policy for carcinogens has assumed that a threshold does not exist for a carcinogenic substance. Public policy for carcinogens in the U.S. builds the concept of no-threshold into risk assessment approaches. More about the idea of risk assessment will be developed in Chapter 2.

Risk assessment and the concept of extrapolating from high dose to low dose are keys to understanding the issues and conflict surrounding asbestos in a nonoccupational setting. The concept of the TLV was based on the belief that the working environment could be controlled, and the likelihood of injury reduced to a minimum by setting threshold values to protect against asbestosis disease. The values needed to be constantly reviewed so they would not be "frozen" by time and use and would take into account new and more accurate data on other manifestations of asbestos disease. In the case of asbestos, "freezing" values sadly became a reality. As scientists revealed the carcinogenic nature of asbestos, the TLV developed to protect against asbestosis proved inadequate to protect against cancer. Indeed it was subsequently demonstrated that the 5 mppcf TLV did not protect against abestosis. The TLV of 5 mppcf remained a recommended guideline, not a legally-enforceable standard, until the passage of the Occupational Safety and Health Act of 1970 required federal enforceable standards. In the meantime, concern about asbestos in the workplace along with the inadequacy of the TLV for asbestos developed, both because of increased awareness of the carcinogenic properties of inhaled asbestos and the increasing utilization of asbestos.

Under present health and safety regulatory procedures for industry in the United States, the Occupational Safety and Health Administration's (OSHA) permanent health standards for air contaminants require an initial determination of concentrations of the agent in air. OSHA has regulated asbestos since 1971. Before 1971, only the Walsh Healy Public Contracts Act of 1958 for contractors with the federal government contained federally-mandated exposure standards, and a few states established their own values. In May 1971, OSHA's initial promulgation of an asbestos in air standard was 12 f/cc of air Permissible Exposure Limit (PEL). The limits were intended primarily to protect against asbestosis, offering only a limited degree of cancer protection. Since 1971, the standard has periodically been revised downward. The agency believed, "sufficient medical and scientific evidence had been accumulated to warrant the designation of asbestos as a human carcinogen," and that "advances in monitoring and protective technology made reexamination of the standard desirable." By 1986, the PEL in the standard had been reduced to 0.2 f/cc. OSHA believed 0.2 f/cc reduced significant risk from exposure, based on *substantial evidence* to be the lowest feasible level. In 1994, OSHA again reduced the PEL, to 0.1 f/cc. The imperative for control of asbestos in the workplace was and is regulation. Table 1 illustrates the change over time in United States Asbestos Standards from 1938 to 1994.[42]

TABLE 1
U.S. Asbestos Standards, 1938–1994

Year	Sponsor	Status	Million particles/cm	Fiber/cc
1938	Dreessen et al.	Recommended TLV	5	30[a]
1946	ACGIH	Adopted TLV	5	30[a]
1970	ACGIH	Adopted TLV	2	12[a]
1971	ACGIH	Proposed TLV	—	5
1971	OSHA	Emergency TWA	—	2
1975	OSHA	Proposed TWA	—	0.5
1976	OSHA	Adopted TWA	—	2
1976	NIOSH	Recommended TWA	—	0.1
1983	OSHA	ETS (TWA)	—	0.5
1984	OSHA	Proposed TWA	—	0.4 or 0.2
1986	OSHA	Adopted TWA	—	0.2
1994	OSHA	—	—	0.1

[a] Approximate fiber equivalent.

Note: ACGIH = American Conference of Governmental Industrial Hygienists; ETS = Emergency Temporary Standard; NIOSH = National Institute for Occupational Safety and Health; OSHA = Occupational Safety and Health Administration; TLV = Threshold Limit Value; TWA = Time-weighted average.

REFERENCES

1. Hunter, D., *The Diseases of Occupations*, Little Brown and Company, Boston, MA, 1955.
2. Hunter, D., *The Disease of Occupations,* Little Brown and Company, Boston, MA, 1955, 1009.
3. Bernarde, M., *Asbestos the Hazardous Fiber*, CRC Press, Boca Raton, FL, 1990, 3.
4. Corn, J., *Response to Occupational Health Hazards,* Van Nostrand Reinhold, New York, 1992, 70.
5. Bernarde, M., *Response to Occupational Health Hazards,* Van Nostrand Reinhold, New York, 1992, 30–40.
6. Corn, J., *Response to Occupational Health Hazards,* Van Nostrand Reinhold, New York, 1992, 91.
7. Corn, J., *Response to Occupational Health Hazards,* Van Nostrand Reinhold, New York, 1992, 90.
8. Bernardi, M., *Asbestos the Hazardous Fiber,* CRC Press, Boca Raton, FL, 1990, 31.

9a. *Report of the Royal Commission on Matters of Health and Safety Arising from the Use of Asbestos in Ontario*, Vol. 1, Ontario Ministry of the Attorney General, Ontario, Toronto, 1984, 75.

9b. National Research Council, *Asbestiform Fibers: nonoccupational health risks*, National Academy Press, Washington, D.C., 1984, 7.

10. *Report of the Royal Commission on Matters of Health and Safety Arising from the Use of Asbestos in Ontario*, Vol. 1, Ontario Ministry of the Attorney General, Ontario, Toronto, 1984, 77.

11. Cullen, M., Controversies in Asbestos-Related Lung Cancer, in *Occupational Medicine, Occupational Pulmonary Disease*, Vol. 2, Rosenstock, L., Ed., Hanley & Belfus, Inc., Philadelphia, PA, 1987.
12. *Report of the Royal Commission on Matters of Health and Safety Arising from the Use of Asbestos in Ontario*, Vol. 1, Ontario Ministry of the Attorney General, Ontario, Toronto, 1984, 86.
13a. Auribault, M., Note sur l'hygiene et la security des ouvriers dans les filatures et tissages d'amante, *Bull de l'Inspection due Travail,* 14, 126, 1906.
13b. Murray, H.M., Statement before the committee in the minutes of evidence, *Report of the Department of Committee on Compensation for Industrial Disease*, His Majesty's Stationary Office, London, 1907.
14. *Report of the Royal Commission on Matters of Health and Safety Arising from the Use of Asbestos in Ontario*, Vol. 1, Ontario Ministry of the Attorney General, Ontario, Toronto, 1984, 98.
15. Craighead, J. and Mossman, B., The pathogenesis of asbestos-associated disease, *New Eng. J. Med.*, 36(24), 1453, 1982.
16. Corn, J. and Corn, M., Changing approaches to assessment of environmental inhalation risk: a case study, *The Milbank Quarterly*, 73, 1, 1995.
17. Ozonoff, D., Failed warnings: asbestos-related disease and industrial medicine, in *The Health and Safety of Workers*, Bayer, R., Ed., Oxford University Press, New York, 1988, 139.
18. Auirbault, M., Note sur l'hygiene et la security des ouvriers dans les filatures et tissages d'amante, *Bull de l'Inspection due Travail,* 14, 126, 1906.
19. Murray, H.M., Statement before the committee in the minutes of evidence, *Report of the Department of Committee on Compensation for Industrial Disease,* His Majesty's Stationary Office, London, 1907.
20. Cooke, W.E., Pulmonary asbestosis, *Br. Med. J.*, 2, 1024, 1927.
21. Pancoast, H.K., Miller, T.G., and Landis, H.R.M., A roentenologic study of the effects of dust inhalation upon the lungs, *Trans. Assoc. Am. Phys.*, 32, 97, 1917.
22. Merewether, E.R.A. and Price, D.W., *Report on the Effects of Asbestos Dust on the Lungs and Dust Supression in the Asbestos Industry*, His Majesty's Stationary Office, London, 1930.
23. U.K. Asbestos Industry Regulations, *Statutory Rules and Orders #1140*, His Majesty's Stationary Office, London, 1931.
24. Hoffman, F.L., *Mortality from Respiratory Diseases in the Dusty Trades*, U.S. Bureau of Labor Statistics, Bull. #231, Washington, D.C., 1918.
25. Lynch, K.M. and Smith, W.A., Asbestos bodies in sputum and lung, *JAMA,* 95, 659, 1930.
26. Donnelley, J., Pulmonary asbestosis, *AJPH,* 23, 1275, 1933.
27. Elllman, P., Pulmonary asbestosis: its clinical, radiological and pathological features and associated risk of tuberculosis infection, *J. Ind. Hyg.*, 15, 165, 1933.
28. Lanza, A.J., McConnell, W.J., and Fehnel, J.W., Effects of the inhalation of asbestos dust on the lungs of asbestos workers, *Public Health Reports,* 50, 1, 1935.
29. Dreessen, C.W., A study of asbestos in the asbestos textile industry, *U.S. Public Health Bull. #241*, Washington Government Printing Office, 1938.
30a. Lanza, A.J., Ed., *Silicosis and Asbestosis*, Oxford University Press, New York, 1938.
30b. Clark, W.I. and Drinker, P., *Industrial Medicine*, National Medical Book Company, New York, 1935.
30c. Rosenau, M.J., *Preventive Medicine and Hygiene*, Appleton-Century-Crofts, New York, 1935.

31. Lynch, K.M. and Smith, W.A., Pulmonary asbestosis III: carcinoma of the lung in asbestos-silicosis, *Am. J. Cancer*, 24, 56, 1935.
32. Doll, R., Mortality from lung cancer in asbestos workers, *Br. J. Indust. Med.,* 12, 81, 1955.
33. Lynch, K.M. and Smith, W.A., Pulmonary asbestosis III: carcinoma of the lung in asbestos-silicosis, *Am. J. Cancer*, 24, 56, 1935.
34. Gloyne, S.R., Two Cases of Squamous Carcinoma of the Lung Occurring in Asbestosis, *Tubercle,* 17, 4, 1935.
35. Doll, R., Mortality from lung cancer in asbestos workers, *Br. J. Indust. Med.,* 12, 81, 1955.
36. Annals of the New York Academy of Sciences, *Biological Effects of Asbestos*, Vol. 32, 1965.
37. Corn, J., *Response to Occupational Health Hazards,* Van Nostrand Reinhold, New York, 1992, 89.
38. Wagner, J.C., Sleggs, C.A., and Marchand, P., Diffuse pleural mesothelioma and asbestos exposure in the north western cape province, *Br. J. Ind. Med.*, 17, 260, 1960.
39a. Castleman, B.I., *Asbestos Medical and Legal Aspects*, Harcourt Brace Jovanovich Publications, New York, 1964.
39b. Lilienfeld, D.E., The silence: the asbestos industry and early occupational cancer research — a case study, *Am. J. Publ. Health*, 81(6), 791, 1991.
40. Corn, J.K., *Protecting the Health of Workers: The American Conference of Governmental Industrial Hygienists 1938–1988*, ACGIH, Inc., Cincinnati, OH, 1989, 59.
41. Dreessen, C.W., Dallavalle, J.M., Edwards, T.I., Miller, J.W., and Sayers, R.R., A study of asbestosis in the asbestos textile industry, *Publ. Health Bull.,* 241, Washington, D.C., United States Public Health Service.
42. Corn, J.K. and Corn, M., Changing approaches to assessment of environmental inhalation risk: a case study, *The Milbank Quarterly*, 73, 99, 1995.

2 From Occupational to Non-occupational Exposure

AWARENESS OF NON-OCCUPATIONAL EXPOSURES

Until the late 1970s, almost all interest in adverse health effects due to asbestos inhalation centered on effects of relatively high levels of occupational exposures. By the end of the decade new interest in health effects of asbestos focused on whether the general public was at risk from asbestos in commercial buildings, schools, and homes. The growing awareness of the hazards of asbestos in the workplace suggested a need to scrutinize other possible areas of asbestos exposure not covered by the Occupational Safety and Health Act, namely non-occupational exposure.

Uneasiness arose about the possible untoward effects on families living in the vicinity of an asbestos textile mill in Paterson, New Jersey. Air laden with asbestos exhausted from a mill located in a residential neighborhood caused concern that residents who did not work at the factory might be exposed to asbestos. Researchers who studied the area reported the following findings at a New York Academy of Sciences meeting in 1979. They inspected homes downwind of the textile mill and found evidence of amosite in the attics of homes around Paterson (Riverside area). They then compared causes of death and death certificates with residents of Totowa, an area miles away from Riverside and comparable in size. Totowa had normal ambient levels of airborne asbestos, i.e., levels lower than Riverside, the area closer to the mill. Totowa also had higher disease rates than Riverside. The researchers concluded that indirect or inadvertent contact with asbestos did not appear to impose additional risk on the residents of Paterson.[1] However, cases of mesothelioma did occur in non-occupationally exposed individuals who lived in neighborhoods with industrial sources of asbestos. Wagner, Sleggs, and Marchand indicated existence of non-occupational mesothelioma in mining areas of Northwest Cape Province in South Africa. They showed that of 33 mesothelioma cases reported during a 5-year period, approximately half were occupational. All but one of the others resulted from non-occupational exposures, i.e., living or working in the area of mining activity.[2]

Newhouse and Thompson investigated the occupational and residential background of 76 individuals who died of mesothelioma in a London hospital. They found that 45 had worked in an asbestos industry, 9 of the remaining 31 lived with a person who was employed in an asbestos industry, and 11 lived within half a mile of an asbestos factory.[3] Bohlig and Hain documented environmental exposure near a factory in Hamburg.[4] Harries showed a risk of asbestos disease from indirect occupational exposure of men in the shipbuilding industry who worked near others who either placed or removed asbestos.[5] A 1976 review of 34 cases of mesothelioma from household contact with asbestos in 9 countries reported 4 new cases among family members of 1,664 asbestos workers.[6] Studies of geographical distribution of cases of mesothelioma in the United Kingdom over a 10-year period indicated that nearly all new cases of mesothelioma came from areas with recognized sources of asbestos.[7] This research presented evidence that indeed a health hazard did exist from community exposure to ambient asbestos. Along with knowledge of community exposure the awareness of hazards of asbestos in the workplace suggested that areas of asbestos exposure, other than occupational, might exist. Thus, extrapolation from occupational settings to non-occupational settings that contained asbestos levels orders of magnitude less than those in the workplace had begun. By the end of the 1970s, at the same time that asbestos in schools became a target of concern, asbestos workers of the 1940s and 1950s appeared with large numbers of cases of lung cancer, asbestosis, and mesotheliomas.[8] The personal injury litigation cases resulting from occupational exposure to asbestos were highly publicized.[9]

By the 1980s, asbestos in schools had become a compelling issue in spite of the fact that no case of asbestos-related disease attributed to occupancy of a school or other public building had been reported.[10] The well-being of building occupants became an environmental issue engendering fear and leading to public policy decisions with far reaching ramifications.

In the 1970s, emerging public knowledge about health risks associated with asbestos in the workplace, and the excessive number of occupational illnesses that came to light, sparked efforts to control them. The Occupational Safety and Health Administration (OSHA) promulgated the first federal enforceable standard for asbestos in the workplace in 1972. Constant lowering of federal permissible exposure limits has occurred since then. The growing knowledge about health risks associated with asbestos in the workplace coupled with the realization that the "magic mineral" had been utilized in building materials since the 1940s and in an estimated 3,000 products (i.e., insulation, floor tiles, piping, roof shingles, linoleum, electric wire casing, and reinforced concrete), some aging and beginning to crumble, shifted interest from workers to building occupants. Reports of mesothelioma from household exposures, finding of asbestos bodies in lungs, at autopsy, of many city dwellers, asbestos fibers found in the environment from lagging of high rise buildings, and the involvement of the Environmental Protection Agency (EPA) with the Clean Air Act of 1970 all focused attention on non-occupational exposure to asbestos.

RISK ASSESSMENT

Before proceeding further it is necessary to present some important technical concepts, seldom explained to the public or to Congress, and thus poorly understood. They were underutilized or even ignored by EPA, the regulatory agency responsible for environmental policy. The concepts include: threshold, extrapolation from high dose to low dose, and risk assessment.

The critical idea of a threshold is that a level of exposure to a toxic substance exists below which no adverse health effect will occur. It is based on the concept of a dose-response relationship for a population of people or laboratory animals. It involves measurement and employs two principles. First, a systematic dose-response relationship exists between severity of exposure to a hazard and the degree of response in the population exposed. Second, as the exposure level decreases, there is a gradual reduction in the occurrence of injury. The risk of injury becomes negligible when the level falls below certain acceptable levels. The idea of threshold limits was developed by toxicologists and industrial hygienists who understood that the "dose makes the poison."[11] The toxic effects of high exposures to asbestos fibers in the workplace were well-documented, especially for asbestosis which develops after years of intense, high exposures. Health risks associated with breathing large amounts of asbestos (high dose) fibers were and still remain clear and well-documented. The dose-response curve, i.e., the amount of fiber inhaled and the related amount of disease at exposures less than "high," is less certain.

Conflict over asbestos and disease focuses on whether there is a threshold level for asbestos below which no adverse effects occur. One school of thought believes that no safe level exists. Others contend that a threshold does exist. This idea is a key one to understanding the conflict over asbestos and disease in the 1970s and 1980s. In the United States, there was a consensus that a threshold existed for asbestos, but the assumption was made for regulatory purposes that a threshold for carcinogens did not exist. The public remained unaware that there were conflicting schools of thought on the subject.

Studies of workers exposed to excessive amounts of asbestos in industry, often years earlier, presented the only available data to relate the amount of exposure to the risk of cancer. Because they lacked data on non-occupational, low-level exposure to asbestos, scientists extrapolated risks associated with past-high industrial exposure levels to obtain hypothetical risks at much lower environmental exposures. They extrapolated from high dose to low dose employing a linear, no-threshold model for lung cancer, which assumed that any asbestos exposure, no matter how small or for how short a duration of time, will result in increased risk of cancer. According to Weill and Hughes, "This mode necessarily results in an estimate of lung cancer cases attributable to any specific asbestos exposure. There is general agreement that the model possibly overstates the potential risk from low cumulative asbestos levels but in all probability does not underestimate it."[12] Remember, risk is related to exposure.

Cullen discussed the question of whether a threshold dose or an exposure beneath which asbestos is noncarcinogenic exist. He presented arguments for and against a threshold, noting that each argument suffers from a lack of substantial detail. "The fundamental evidence for a threshold is the absence of a significant excess of lung cancer at lowest doses in most published mortality studies. On the other hand, data refuting a threshold effect are equally unsatisfactory."[13] Cullen said that from available information certain conclusions can be reached. "On the one hand analysis of both animal and human epidemiologic data does demonstrate that if there is an increased risk associated with "low" levels of asbestos, it surely is not nearly as great as that seen at higher exposure levels in any given exposure setting; so we may be confident that there is no exaggerated effect (compared to a linear dose response, for example) at "low" levels. Conversely, the clearly positive risks demonstrated in two studies with very low levels of exposure or brief duration of exposure make prediction of any particular threshold dose highly dubious with present data."[14] Cullen summed up by saying that the best estimate for the response to a low level is linearity, as argued by Peto[15] and McDonald,[16] based on higher doses, biological inference, and the utilitarian view that this estimate would lead to the most reasonable public health approach. Clearly, uncertainties did abound. Nevertheless, decision makers and the public were not apprised of these ambiguities.[17] Apparently neither Congress, nor the general public understood these concepts, knew of their utility, or were informed about the uncertainties inherent in extrapolation from high to low dose. Policy decisions as well as research decisions, because of scientific uncertainties, were fast becoming social and political phenomenon in the 1970s and early 1980s. It is interesting to note that advances in genetic science, including evidence of DNA repair, have intensified the threshold debate in the 1990s.[18]

Concomitant with the debate on thresholds for carcinogens, the framework of risk assessment was being developed in an attempt to create a framework for regulatory decision making in order to set priorities in the presence of a growing list of environmental concerns. Although certainly not a new idea, risk assessment emerged coincidentally with formation of the two regulatory agencies, OSHA and EPA. Controversy over policy and regulatory decisions often, and still does, focus on selection of priorities for what hazard to control and the degree of control. Contemporary concern about risk centers on health risks caused by industrial processes rather than natural phenomena such as earthquakes and hurricanes. We know human beings have always had to deal with risk, often defined as the potential for harm, the unwanted negative consequence of an event, or the probability of something harmful happening. Although not a new idea, it recently has been redefined and refined. Covello and Mumpower reviewed the history of risk analysis and risk management in an effort to provide historical perspective to risk assessment. Table 1, "Changes from Past to Present Relevant to Risk Analysis," is based on their review article.[19]

The concept of risk assessment needs explanation if one is to understand or place in historical perspective United States policy for non-occupational exposure to asbestos. Control and prevention of environmental hazards is in large part an exercise in identifying pathways by which a particular agent causes health or environmental damage. Risk assessment permits the assignment of risks to the different pathways, thus becoming a powerful tool for decision making. Because public health decisions are made within a scientific and social framework, addressing risk involves two extremely different kinds of activity: assessing risk — an empirical, scientific activity, and judging risk — a normative political activity. The first implies objectivity, while the second implies societal values. In the case of non-occupational asbestos exposure, objectivity fell short of the mark because the EPA never presented a risk assessment for the concern.

Usually considered synonymous with quantitative risk assessment (QRA), risk assessment is defined as the characterization of the potential adverse health effects of human exposures to environmental hazards. It includes:

TABLE 1
Changes from Past to Present Relevant to Risk Analysis

Change	Example
1. Shift in the nature of risk	* From infections to chronic disease * Growth in number of auto accidents
2. Increase in life expectancy	* Female born in U.S. 1900 Life expentancy: 51 * Female born in U.S. 1975 Life expectancy: 75 * Male born in U.S. 1900 Life expectancy: 48 * Male born in U.S. 1975 Life expectancy: 66
3. Increase in new risks	* Chemicals * Radio-active wastes * Pesticides * Nuclear accidents
4. Increase in scientists' ability to measure and identify risks	* Advances in laboratory testing * Epidemiological methods * Computer simulations
5. Increase in the number of scientists and analysts whose work is focused on health, safety, and environmental risks	* Risk assessment, analysis is emerging as an identifiable discipline
6. Increase in the number of formal quantitative risk analyses produced and used	* Use of highly technical quantitative tools
7. Increase in role of federal government in assessing and managing risks	* Increase in number of health, safety, and environmental laws * Increase in number of federal agencies charged with managing health, safety, and environmental risks
8. Increase in participation of special interest groups	* Risk assessment has become increasingly politicized
9. Increase in public interest, concern, and demands for protection	

1. a description of potential adverse health effects based on evaluation of results of epidemiologic, toxicologic, clinical, and environmental research;
2. extrapolation from these results to predict the type, and estimate the extent of health effects in humans under given conditions of exposure;
3. judgments as to the number and characteristics of persons exposed at various intensities and durations; and
4. summary judgments on the existence and overall magnitude of the public health problem, and characterization of the uncertainties inherent in the process of inferring risk.[20]

Table 2, "Elements of Risk Assessment and Risk Management," is from a 1983 document published by the National Academy of Science and is often referred to as the risk paradigm. It illustrates the elements of risk assessment and risk management.[21]

A risk assessment starts with the common concept that the probability of human harm from a toxic substance is a function of two variables, the potency and the dose.* Risk assessors need to develop a method, when assessing the magnitude of a risk, to relate the dose-response curve (information about incidence of disease produced by different doses of a substance) and the exposures that humans incur or are likely to incur. For this end they employ a variety of different mathematical models to relate dose-response to exposure.

Critics of risk assessment assert that: (1) predicting risks by using different mathematical models presents a wide disparity in risks; (2) empirical verification of models is often missing; and (3) uncertainties are inherent in extrapolating from animals to humans and from high dose to low dose. Supporters of risk assessment acknowledge the difficulties but defend risk assessment as an important tool enabling an approach to setting priorities in a rational manner. Supporters of quantitative risk assessment believe it can structure decisions which involve uncertainties, and that efforts to reduce human exposures to the toxic substances require a method to estimate the magnitude of health consequences of alternative control methods.

In less than two decades assessing health risks of non-occupational asbestos presented numerous challenges and became a contentious, devisive issue. The problems arose partly because of the uncertainties about the degree and nature of exposure and also because of the inability to utilize risk assessment and to separate scientific inputs (risk assessment) from political inputs (risk management). Risk assessment is complex. It includes legal, political, ethical, social, and economic components often derived from, and always allied to, value considerations far beyond science.

Risk has been further subdivided into risk management and risk communication. The process of developing policies based on a quantitative risk assessment is referred to as risk management. One reason the policy developed for non-occupational exposure to asbestos, specifically asbestos in schools, floundered and in the judgment of many failed and brought about unintended consequences, was that the EPA did not utilize a risk assessment as a basis for managing asbestos in schools. Regulatory agencies perform risk management under a variety of legislative mandates. Risk management requires the regulatory agency to develop, analyze, and compare options and to select appropriate regulatory responses to a potential hazard.[22] Risk management can and does become politicized because it involves value judgements on issues such as acceptability of risk and reasonableness of cost.

Risk communication is the exchange of information about risk among scientists, decision makers, and the public. The three R's, risk assessment, risk management, and risk communication are often referred to as risk analysis. They can be viewed as a response to controversial and complex environmental issues which require prudent public health policies to regulate environmental risks not negotiable by individuals, and therefore in need of governmental intervention. The concept of risk

* Dose is taken as the exposure multiplied by the time of exposure.

TABLE 2
Elements of Risk Assessment and Risk Management

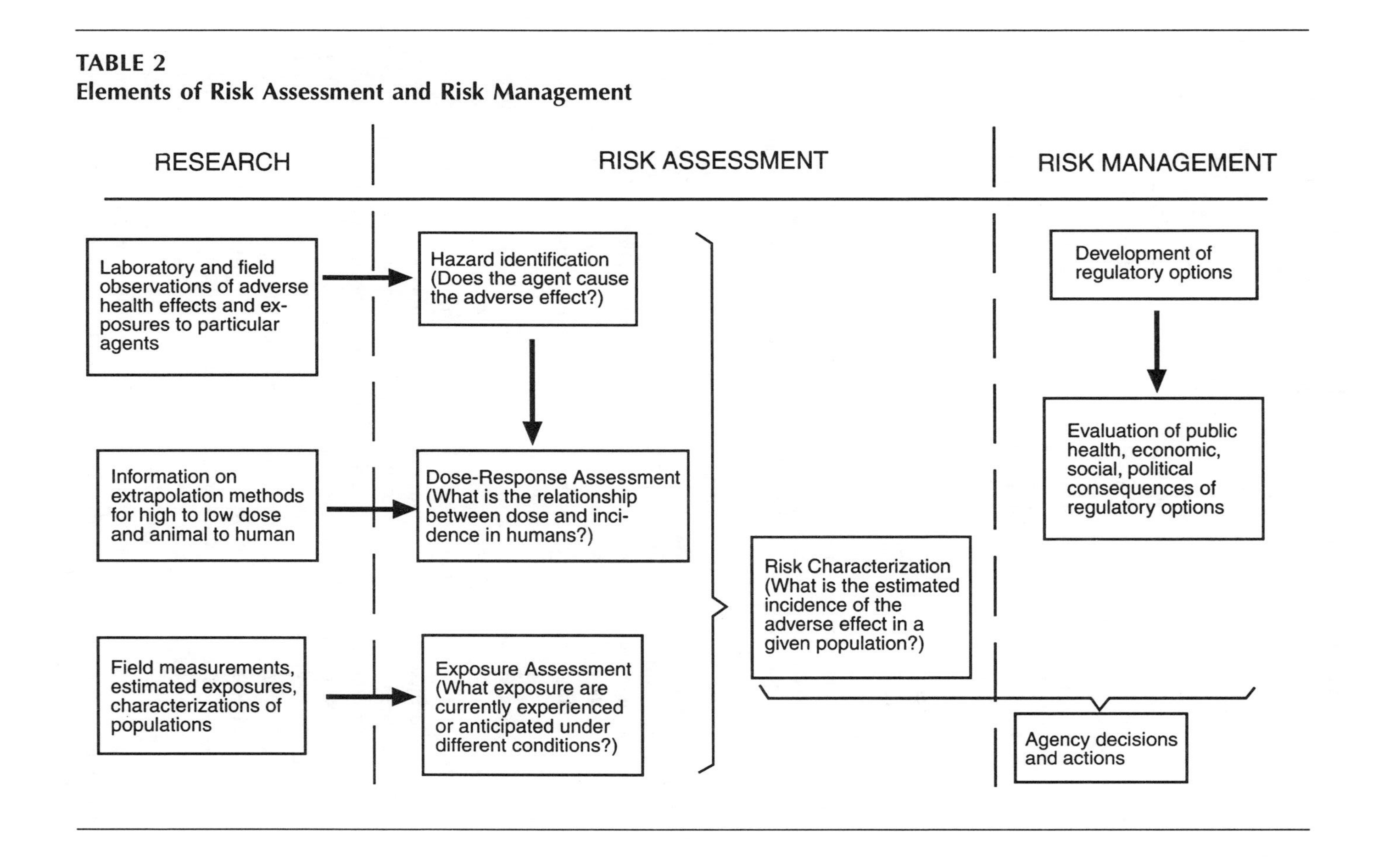

can be traced to biblical times. Our current risk assessment concepts have roots in the early 20th century and in the recognition that occupational disease and exposure to hazards in the workplace, such as lead, mercury, and silica, are related. Early acceptance of the premise that health risk is related to the degree of exposure and to the degree of toxicity of the material, characterized the general approach to risk assessment. Later, risk assessment evolved to include epidemiological data of worker populations and dose-response (toxicological) data that involved animal tests.

The extraordinary growth of the chemical industry after World War II and the resulting proliferation of new toxic hazards led to identification and interest in environmentally caused disease. At the same time toxicological sciences emerged as a tool to assess chemical risk.[23] The Food and Drug Administration (FDA) began to identify chemicals that could be safely added to food and drugs. In 1962, Rachel Carson's book *Silent Spring* focused popular attention and scientific interest on environmental pollution caused by the widespread use of chemicals.[24]

RISK AND ENVIRONMENTAL REGULATION

In the 1960s and 1970s federal environmental laws dealt with toxins and carcinogens. Many of the laws are rooted in the belief that most concerns are environmental in origin and thus can be prevented by reducing human exposure to carcinogenic substances in the environment. Table 3 indicates major federal environmental legislation from 1938 to 1986. Note that most laws were passed within an 11-year period, 1969 to 1980. The legislation encompassed water, air, pesticides, consumer protection, and occupational safety and health. In the 1970s, orientation of environmental policy shifted from air, water, and pesticides to human health.

Debates about environmental regulation quickly created conflict. Divergent views and values led to an inability to separate scientific and political inputs during the decision-making process. Science and policy became intricately entwined and difficult to separate. Much of the controversy arose over the magnitude and management of environmental risks. Controversy developed because of the lack of guidelines for testing and evaluating experimental data and over choice of substances needing regulation. Scientific uncertainties added to the list of problems. To make the situation even more difficult, proponents and opponents of specific policies distorted science to achieve their own objectives.[25] Although many of the problems associated with developing environmental policy could be attributed to the limitations of science, the pervasive persistent issue in regulatory legislation, that of the scientific basis for risk decisions, combined with the controversy over goals and how to achieve them, became controlling factors driving American environmental policy. Policy decisions made for asbestos in schools were arrived at in this context. When the United States embarked on its ambitious road to regulate environmental risks in order to reduce them, it became clear that regulatory issues were intimately associated with subjective aspects of the political process. Policy and decision makers required a method to select priorities. Both the scientific and policy communities responded with a set of techniques they called risk assessment. But it would take time before risk assessment gained acceptance as a tool for regulatory decision making.

TABLE 3
Federal Environmental Legislation

Law		Year Enacted
FDCA	Federal Drug and Cosmetics Act	1938
FIFRA	Federal Insecticide, Fungicide, & Rodenticide Act	1948, 1972, 1975
FHSA	Federal Hazardous Substances Act	1966
NEPA	National Environmental Protection Act	1969
PPPA	Poisonous Packaging Prevention Act	1979
OSHA	Occupational Safety and Health Act	1970
CAA	Clean Air Act	1970, 1977
FWPCA	Federal Water Pollution Control Act (now Clean Water)	1972, 1977
MPRSA	Marine Protection, Research and Sanctuaries Act	1972
CPSA	Consumer Product Safety Act	1972
FEPCA	Federal Environmental Pollution Control Act	1972
SDWA	Safe Drinking Water Act	1974, 1977
HMTA	Hazardous Materials Transportation Act	1974
RCRA	Resource Conservation Recovery Act	1976, 1979
TSCA	Toxic Substance Control Act	1976
SMCRA	Surface Mine Control and Reclamation Act	1977
UMTCA	Uranium Mill Tailings Control Act	1978
CERCLA	Comprehensive Environmental Response, Compensation and Liability Act	1980
SARA	Superfund Amendments and Reauthorization Act	1986
AHERA	Asbestos Hazard Emergency Response Act	1986

It would be difficult to understand the history of American policy for asbestos in school buildings without comprehending the nature and significance of risk assessment. The many problems associated with decisions made about asbestos in school buildings could be attributed to both the capabilities and limitations of risk assessment. Although hardly articulated just 25 years ago, today risk assessment is noteworthy. As noted earlier, contemporary risk assessment in the federal government, utilized for environmental policy, took hold in the 1970s and 1980s. The movement toward more clearly defining health risks began in the 1970s when federal regulatory agencies moved toward utilization and acceptance of risk assessment as an analytical tool for regulatory decision making signaling perhaps one of the most far reaching changes in the United States policy to regulate the environment. By the 1980s, most of the environmental community had come to terms with the concepts of environmental risk assessment. In 1983, William Ruckelshaus, then administrator of EPA, wrote an editorial in *Science* about the need for a government-wide process to assess and manage environmental risks. Ruckelshaus wanted to strengthen risk assessment capabilities in EPA. He wrote, "We need more research on the health effects of substances we regulate. I intend to do everything in my power to make clear the importance of this scientific analysis at EPA. Given the necessity of acting in the face of enormous scientific uncertainties, it is more important than ever that our scientific analysis be rigorous and the quality of our data be high. We must take pains not to mislead people about the risks to their health. We can help to avoid confusion by ensuring both the quality of our science and the clarity of our language in explaining hazards."[26] Ruckelshaus wrote this article in response to what he termed a "troubled emotional period for pollution control," when communities gripped with fear and worry about their health responded emotionally and without facts. Emotionalism dominated policy decisions and the public knew little or nothing of the uncertainties involved in the estimates of risk. Ruckelshaus said that standards set by EPA would from then on, whether technology- or health-related, have a sound scientific base. Sadly this did not occur in the case of asbestos in schools, where emotion overruled science. Ruckelshaus' speech proved to be mere rhetoric, as far as asbestos in schools was concerned.

Nevertheless, it is important to point out some noteworthy events that signalled the movement in the federal government toward risk assessment, because they profoundly affect both the outlook of regulators and thus American environmental policy. They include: (1) the publication of an influential book written by William Lowrance, *Of Acceptable Risk*;[27] (2) the Supreme Court decision which invalidated OSHA's benzene standard for not defining "significant risk;"[28] and (3) the publication of a National Academy of Science report, *Risk Assessment in the Federal Government: Managing The Process.*[29]

In 1976, understanding that a major policy challenge faced regulators and decision makers struggling with the problem of how to evaluate risk, William Lowrance defined risk for regulators, scientists, decision makers, and the public. He pointed out the central role of risk assessment as a tool for organizing and analyzing information, its scientific basis and its limitations. Lowrance posed the following questions:

> How did we determine how hazardous these things are? Why is it that cyclamates one day dominate the market as the principal calorie-cutting sweetener in millions of cans of diet drinks, only to be banned the next day because there is a 'very slight chance' they may cause cancer? Why is it that one group of eminent experts says that medical X-rays (or food preservatives, or contraceptive pills) are safe and ought to be used more widely, while another group of authorities, equally reputable, urges that exposure to the same things should be restricted because they are unsafe? At what point do debates such as that over DDT stop being scientific and objective and start being political and subjective? How can anyone gauge the public's willingness to accept risks? Why must there be these endless controversies over such things as lead, whose effects on health have been known in detail for years? Are people being irresponsible, or is there something about these problems that just naturally spawns confusion? Just what sort of a decision making tool is this notion of safety?[30]

He then explored the problems, the underlying concept of safety itself, and the general features of the social context within which safety decisions are made. He also emphasized contributions of scientists and technically trained people.

Lowrance differentiated between the evaluation or estimate of risk, which is a scientific endeavor, and the acceptance of levels of risk, which involves sociopolitical decision making. Addressing risk therefore involves two extremely different kinds of activity. First, measuring risk, involves measuring the probability and severity of harm. It is an empirical, scientific activity and implies objectivity. Second, judging risk, involves judging the acceptability of risk. It is a normative political activity and implies value. Lowrance wrote: "Safety is not measured. Only when those risks are weighed on the balance of social values can safety be judged: a thing is safe if its attendant risks are judged to be acceptable."[31]

The mixing of the two activities, the assessment of risk and the judgement of risk, in United States environmental standard setting since 1971 had caused much bitterness among identifiable groups associated with standard setting.[32] Often, as was the case with non-occupational asbestos, because there was insufficient scientific data to assess risk, final recommendations reflected judgment of the acceptability of risk and not a combination of measurement and judgment. Lowrance's philosophy became the underpinning for risk assessment in the federal government.

In February 1978, OSHA issued a permanent standard for benzene that lowered the previous occupational permissible exposure limit adopted in 1971 from 10 parts per million (PPM) to 1 PPM, with a ceiling of 15 PPM during any 15 min period.[33] OSHA adopted the stringent 1 PPM based on its "lowest feasible level" policy. There was no animal data or epidemiological data for benzene that indicated human beings contracted leukemia at concentrations significantly higher than the prior level of 10 PPM. OSHA concluded that it wasn't possible to demonstrate a threshold level for benzene-induced carcinogenicity or to establish a safe level for benzene exposure. They therefore decided that the permissible exposure level for benzene should be reduced to the lowest feasible level. They concluded that higher exposures carry greater risk; and since they could not make a determination of what constitutes a safe level of exposure to benzene and that presents no hazard, they could not answer the question of whether a safe level of exposure to benzene exists.[34] OSHA said prudent public health policy required they limit exposure to the maximum extent feasible.

The permanent standard on benzene was challenged in court by the American Petroleum Institute (API) and vacated. In 1980, the United States Supreme Court upheld the lower court's decision to vacate the standard. Although the Court was divided, the plurality concluded that OSHA must show that the toxic substance at issue (benzene) created a "significant" health risk in the workplace and that the lower standard would eliminate or reduce that risk. They said OSHA had failed to make a valid determination that reducing the permissible exposure level of benzene from 10 PPM to 1 PPM is reasonably necessary to protect workers from a risk of leukemia.[35]

A number of aspects of this decision need to be emphasized. First, when the plurality struck down OSHA's policy of regulating carcinogens to the lowest feasible level they said the policy was based on assumptions rather than evidence. Second, the plurality held that an agency was required to show by substantial evidence that "at least more likely than not" the new standard will eliminate or reduce a significant risk in the workplace. Third, the plurality said that the risk from a substance must be quantified sufficiently to enable the Secretary to characterize it as significant in an understandable way.

The bottom line is that in the post-benzene decision era the test of significant risk and utilization of quantitative risk assessment are now part of the process of setting standards to regulate toxic chemicals. The Supreme Court decision profoundly affected future regulatory policy. The "Benzene Decision" reflected the need for risk assessment as a method to select priorities.

The third event that affected the development of risk assessment was the publication in 1983 of a National Academy of Science report entitled, *Risk Assessment in the Federal Government: Managing the Process*. The report suggested that risk assessment and the choice of regulatory options be distinguished from each other. The committee recommended that, "… the regulatory agencies take steps to establish and maintain a clear conceptual distinction between assessment of risks and consideration of risk management alternatives; that is, the scientific findings and policy judgments embodied in risk assessment should be explicitly distinguished from the political, economic, and technical considerations that influence the design and choice of regulatory strategies." They also recommended the development of uniform inference guidelines for the use of federal agencies in the risk assessment process and that risk assessment methods be established for regulatory agencies.[36]

Other factors that should not be overlooked which led to incorporation of risk assessment in regulatory decisions include:

1. New methods for risk assessment began to appear in the scientific literature. There was a significant growth in this area of scientific inquiry in the 1970s.
2. Publication of a Federal Interagency Regulatory Liaison Group document on the subject of risk assessment 1979.
3. Publication of the Science and Technology Policy document on the subject of Risk Assessment 1985.
4. Publication of EPA document, *Risk Assessment and Management: Framework for Decision Making* (1984): EPA Administrator William D. Ruckelshaus embraced the idea of risk assessment.
5. Executive Order 12044 (1978) required regulatory agencies to perform an analysis for regulations that would have significant economic impacts.
6. Executive Order 12291 (1981) replaced 12044 and required all executive agencies to prepare a regulatory impact analysis to help select regulatory approaches in which potential benefits to society from regulation would outweigh potential costs.

Economic impact statements were required and detailed regulatory analysis. These federal guidelines were aimed at "regulatory reform."

Although progress has been made in risk assessment during the past 25 years, many problems remain unsolved, especially how to create a bridge of understanding between risk assessors and policy makers concerning types, levels, distribution, and acceptability of risk.

RISK ASSESSMENT, EPA, AND ASBESTOS IN SCHOOLS

Health risk assessment can provide a systematic approach to the evaluation and estimation of risk to human health in order to guide the formulation of environmental policy and regulatory decisions and to link health risk research to decision making. The Office of Technology Assessment (OTA) stressed the interdependency of research activities, the risk assessment process, and policy making in a 1993 document.[37] Unfortunately, this did not occur in 1979 when federal concern about health risks for people who lived or worked in buildings containing asbestos commenced. EPA, the main federal agency responsible for asbestos in school activity, did not fully utilize its own risk assessments to place either potential or real risks in perspective for the Congress or the public. The agency did not fully come to grips with these ideas when developing its policy for non-occupational exposure to asbestos. They often went through the motions of risk assessment and initiated guidance documents without environmental measurements of asbestos exposure levels that might cause health risks to building occupants. Table 4 indicates the risk assessment documents produced for EPA between 1980 and 1986. EPA did not utilize or internalize many of these documents until after 1990. At that time the agency literally walked away from the mess they had made.

Table 4 displays salient characteristics of four quantitative risk assessments for projected deaths from asbestos in schools. The two EPA assessments were provided for the public. The Hughes and Weill projections for school children were based on six years of exposure, not a lifetime of exposure, as were the EPA assessments. The results do not greatly differ, suggesting the uncertainties in these estimates, stemming from differing assumptions by the risk assessors, did not materially affect results. The Health Effects Institute lifetime estimates of deaths for children and building occupants does not differ greatly from the EPA estimates. In 1991, exposure data for maintenance and repair workers were sparse. More extensive data are available today, but it is noteworthy that the HEI 200/100,000 estimate in Table 4 is less than the OSHA 340/100,000 estimate for lifetime exposure at the current Permissible Exposure Limit (PEL) of 0.1 fiber/cm^3 greater than 5 μm in length for working men and women, 45 years, 50 weeks/year, 40 h/week exposure to asbestos. Thus, building repair and maintenance workers are in compliance with the current legal exposure limit.

The consistency of the low risk estimates for school children and building occupants in these four risk assessments, spanning the time period 1980 to 1991, is not consistent with the emergency approach to asbestos by EPA. The agency should set priorities based on risk. Those estimates of risk should have placed low priority on an all out program for abatement of asbestos in schools. The question remains: Why didn't EPA act in accordance with its own risk estimates?

In 1980, EPA applied a linear dose-response curve to results of a study of asbestos insulation workers exposed to high doses of asbestos to estimate the risk presented by low-level asbestos exposure in schools. Using fragmentary building exposure data collected in France, EPA concluded that in a worst case situation, between 100 and 6,800 premature cancer deaths will result from exposure to prevailing asbestos levels in schools. EPA then developed a "most reasonable" estimate of 1,100 premature deaths.[38] Reviewers criticized EPA's analysis for overstating the risk presented by asbestos in schools.[39] In fact, EPA asbestos policy decisions did not link research about risks from measured non-occupational, low levels of exposure to its decisions.

TABLE 4
Asbestos Risk Assessments (RA)

Performed by	Year	Purpose	Predicted deaths per 100,000[a]	Comments	Reference
U.S. EPA	1980	Assess risk to occupants and to workers in schools to support agency policy for asbestos in schools	0.32 school children 0.64 school teachers, administrators, custodians, and maintenance workers	Lifetime exposure estimated from measurements made in buildings "similar" to schools Exposure estimates adjusted to consider deterioration of ACM	U.S. Environmental Protection Agency, support document asbestos-containing materials in schools, EPA-560/12-80-003, 1980.
U.S. EPA	1986	Serve as the scientific basis for the review and revision of the National Emission Standards for asbestos as a hazardous air pollutant	0.5 females 1.7 males	Exposure estimated from continuous lifetime exposure assumptions of 0.0001 f/ml and 0.01 f/ml	U.S. Environmental Protection Agency, Airborne asbestos health assessment update, EPA/600/8-84/003F, 1986.
J.M. Hughes and H. Weill	1986	Published review of quantitative risk assessments of asbestos health risks	0.5 to 1.5 school children	Exposure estimated as 0.001 f/ml for 6 years	J.M. Hughes and H. Weill, Asbestos exposure — quantitative assessment of risk, *Am. Rev. Respir. Dis.*, 133, 5, 1986.
Health Effects Institute	1991	Respond to Congressional mandate to assess adverse health effects to asbestos in buildings	0.6 school children 0.4 occupants public buildings 200 maintenance and repair workers	Exposure estimates based on all available airborne asbestos data Estimates lifetime risks for occupants in schools and public buildings and for custodians and janitors	Health Effects Institute-Asbestos Research (HEI-AR),Asbestos in public and commercial buildings: a literature review and synthesis of current knowledge, Cambridge, MA, 1991.

[a] Excess deaths due to lung cancer and mesothelioma.

The basis for regulation, health effects, and magnitude of exposure to asbestos were not emphasized by EPA. They stressed epidemiological evidence, based on high-dose to maintain that asbestos in schools causes cancer. In the agency's risk assessment a major study of 12,000 asbestos insulation workers was cited. For example, "Cancers induced by asbestos were responsible for elevating the overall mortality rate by approximately 50 percent. Seventeen percent of all deaths in this group were due to asbestos, pleural mesothelioma and peritoneal mesothelioma."[40] "These causes are virtually unheard of in the general population. The International Agency for Research on Cancer (IARC) has listed asbestos among only 18 chemicals, groups of chemicals and industrial processes for which the evidence of human carcinogenicity is conclusive."[41] The EPA also found that, "adverse health effects of non-occupational exposure to asbestos have been amply demonstrated. Some persons whose known exposure to asbestos has been from living in the same households as asbestos workers or in the neighborhoods of asbestos mines, mills and processing facilities have developed mesothelioma and signs of asbestosis."[42] There was no discussion of dose, although EPA must have extrapolated from high to low dose.

Nevertheless, EPA assumed that the magnitude of exposure to asbestos in buildings with friable asbestos justified regulating asbestos in schools and found it "highly likely that exposure to asbestos in schools may increase the risk of developing numerous types of cancer most notably plural and peritoneal mesothelioma and lung cancer."[43] The risk assessments EPA referred to in the Federal Register addressed human health. They did not address extrapolation from high to low dose, or exposure measurements. For example, they failed to ascertain just how much asbestos was in the ambient air of schools, if epidemiological studies were performed, or if there were cases of school children or teachers exposed to asbestos in school buildings who suffered from either cancer or mesothelioma as a result of exposure to asbestos fibers in schools. With hindsight, EPA's risk assessment appears minimal. Based on its inadequate risk assessment EPA's program to regulate asbestos in school buildings moved forward. The Technical Assistance Program (TAP), begun in 1979, culminated in 1986 with unanimous passage of the Asbestos Hazard Emergency Response Act (AHERA) by the United States Congress.

Other federal agencies already had responsibility to regulate asbestos. OSHA, the agency responsible for worker health and safety, regulated asbestos in the workplace beginning in 1971 when it promulgated the occupational asbestos standard with initial promulgation of an asbestos-in-air standard of 12 f/cc PEL. OSHA steadily revised the PEL downward to a 1986 PEL of 0.2 f/cc and a 1994 PEL of 0.1 f/cc.* Other federal agencies with responsibility for non-occupational asbestos include the Department of Education and the Consumer Product Safety Commission. EPA and the Department of Education are responsible for federal action concerning asbestos in schools. EPA acts under the Clean Air Act and the Toxic Substances Control Act (TSCA). The Department of Education administers the Asbestos School Hazard Detection and Control Act.

In 1971, EPA added asbestos to the list of materials regulated under the Clean Air Act and the National Emission Standard for Hazardous Pollutants (NESHAP).

* See Table 1, Chapter 1.

EPA intended to prevent the release of asbestos fibers into air outside of all buildings. In 1973, EPA determined asbestos a hazardous pollutant and banned the spraying of asbestos-containing insulation in buildings. In 1978, EPA extended the ban to all uses of sprayed-on asbestos on building structures, beams, ceilings, walls, pipes, and conduits. The agency also mandated work practices when demolishing or renovating a building containing asbestos material to minimize the release of asbestos fibers into the outdoor environment.[44] Thus, OSHA retained jurisdiction in the workplace and EPA-regulated emissions of asbestos to the environment outside of buildings.

In 1979, EPA began to target its asbestos regulatory authority to protect school children from adverse health effects of airborne asbestos inside school buildings. In that year, the Office of Toxic Substances (within EPA) coordinated efforts, under TSCA's broad authority, to control chemical substances to protect the public health and environment against unreasonable risk, and launched the Technical Assistance Program (TAP). TAP was a nonregulatory attempt to educate school officials, state government officials, parents of students, and school employees to the potential hazards of airborne asbestos and to encourage voluntary identification and correction of asbestos hazards in schools. EPA published the first in its series of guidance documents, *Asbestos-Containing Materials in School Buildings*,[45] better known as the "Orange Book." The "Blue," "Purple," and "Green" books followed. EPA published numerous guidance documents between 1979 and 1990. The "Orange Book" was mailed to all public and private schools in the United States. It provided information and advice on inspection procedures, sampling, and abatement methods.

EPA proposed the rule entitled, "Friable Asbestos-Containing Material in Schools; Identification and Notification" in 1980. It became effective in 1982, promulgated under TSCA. The rule required local educational agencies (LEAs) to: (1) inspect school buildings for friable materials applied to structural surfaces; (2) take at least three samples of each type of material found; and (3) have samples analyzed by polarized light microscopy (PLM). The rule required LEAs to inform employees, parents, and parent-teacher groups about the presence and location of asbestos in the school, if a school building contained friable asbestos-containing materials (ACM).[46]

The Worker Protection Rule of 1983 was an attempt to address employee protection. It did not become effective until 1986. In 1984, President Reagan signed the Asbestos School Hazard Abatement Act (ASHAA). The act established a program to ascertain the extent of danger to the health of school children and employees caused by ACM in school buildings, provide scientific and technical guidance to states and to LEAs on asbestos abatement and identification, and provide financial assistance to abate asbestos hazards in schools. It authorized $600 million to be spent over a seven year period for loans and grant to LEAs. In 1986, Congress passed and the President signed the Asbestos Hazard Emergency Response Act (AHERA). Rule making for AHERA was completed in 1987. AHERA covered ACM in school buildings. It required inspection by accredited inspectors and identification of asbestos, both friable and nonfriable, by October 12, 1988. Inspection of buildings with asbestos required sampling and assessment, as well as training to provide information on asbestos and how to deal with it. Custodians or maintenance workers needed two hours of training. A management plan had to be submitted and implemented by July 1, 1989. Warning labels were to be placed in maintenance areas, readily visible to people who worked in the area. The law also included accreditation requirements for those in abatement-related positions and enforcement measures.[47]

Federal policy, as we will see, began with a bang and ended with a whimper. Criticism of the policy, in general, and of AHERA specifically reached the public slowly. By 1990, research and new data showed levels of asbestos in schools to be low, indeed, often lower than outdoor air.[48] In 1989, the *New England Journal of Medicine* published a review of asbestos research challenging the conventional wisdom about asbestos. Its authors concluded, in a review of asbestos-related disease, that "determinations of respirability, the airborne concentrations of fibers, friability, fiber size and most important, the type of asbestos (particularly when amphiboles are involved) are prerequisites for any rational determination of the need for asbestos abatement."[49] In 1990, a paper published in *Science* specifically criticized public policy for asbestos.[50] The editor endorsed the article in an editorial entitled, "The Asbestos Removal Fiasco."[51] After the *Science* article and editorial, a number of articles began to appear in the popular media, as well, suggesting that public policy, including past enormous expenses for asbestos abatement, may have been ill-advised, poorly conceived, and certainly not based on adequate scientific assessment of the problem. EPA responded by reversing its policy for asbestos removed from school buildings.

The question remains, how did the asbestos panic begin and what caused it to continue largely unexamined for over ten years and at a projected cost for abatement of 50 to 150 billion dollars?

REFERENCES

1. Hammond, E.C., Garfunkel, L., Selikoff, I.J., and Nicholson, W.N., Mortality experience of residents in the neighborhoods of an asbestos factory, *Ann. N.Y. Acad. Sci.*, 330, 417, 1979.
2. Wagner, J.C., Sleggs, C.A., and Marchand, P., Diffuse pleural mesothelioma and asbestos exposure in the north-western cape province, *Br. J. Ind. Med.*, 17, 260, 1960.
3. Newhouse, M.L. and Thompson, H., Mesothelioma of pleura and peritoneum following exposure to asbestos in the London area, *Br. J. Ind. Med.*, 22, 261, 1966.
4. Bohlig, H. and Hain, E., Cancer in relation to environmental exposure, type of fibre, dose, occupation and duration of exposure, in *Proceedings of the Conference on Biological Effects of Asbestos*, Bogovski, P. et al., Eds., Lyon, France, 1973, 217.
5. Harries, P.G., Experience with asbestos disease and its control in Great Britain's naval dockyards, *Environ. Res.*, 11, 261, 1976.
6. Anderson, H.A., Lillis, R., Daum, S.M., Fischbein, A.S., and Selikoff, I.J., Household contact and asbestos neoplastic risk, *Ann. N.Y. Acad. Sci.*, 271, 311, 1976.

7a. Gilson, J.C., Asbestos health hazards. Recent observations in the United Kingdom, in *Pneumoconiosis Proceedings of the International Conference*, Shapiro, H.A., Ed., Oxford University Press, 1970, 173.

7b. Greenberg, M. and Davies, L., Mesothelioma register 1967–1968, *Br. J. Ind. Med.*, 31, 91, 1974.

8. Corn, J. and Starr, J., Historical perspective on asbestos: policies and protective measures in World War II shipbuilding, *Am. J. Ind. Med.*, 11, 359, 1987.

9. Brodeur, P., *Outrageous Misconduct: The Asbestos Industry on Trial,* Pantheon Books, New York, 1985.
10. Gaensler, E., Asbestos exposure in buildings, *Occupat. Lung Dis.*, 13(2), 19, 232, 1992.
11. Ottobani, *The Dose Makes the Poison,* Vincente Books, Berkeley, CA, 1984.
12. Weill, H. and Hughes, J.M., Asbestos as a public health risk: disease and policy, *Ann. Rev. Publ. Health,* 7, 171, 1986.
13. Cullen, M., Controversies in asbestos-related lung cancer, in *Occupational Medicine — Occupational Pulmonary Disease*, Vol. 2, Rosenstock, L., Ed., 2(2), 265, 1987.
14. Cullen, M., Controversies in asbestos-related lung cancer, in *Occupational Medicine — Occupational Pulmonary Disease*, Vol. 2, Rosenstock, L., Ed., 2(2), 265, 1987.
15. Peto, J., Dose-response relationships for asbestos-related disease: implications for hygiene standards. Part II, *Ann. N.Y. Acad. Sci.,* 330, 195, 1979.
16. MacDonald, J.C., Liddell, F.D.K., Gibbs, G.W. et al., Dust exposure and mortality in chrysotile mining 1910–1975, *Br. J. Ind. Med.,* 37, 11, 1980.
17. Murray, T.H., Regulating asbestos: ethics, politics, and scientific values, *Sci. Tech. Hum. Values,* 11(3), 13, 1986.
18. Mossman, B., Bignon, J., Corn, H., Seaton, A., and Gee, B.L., Asbestos: scientific developments and implications for public policy, *Science,* 247, 1990.
19. Covello, V. and Mumpower, J., Risk analysis and risk management: an historical perspective, *Risk Anal.*, 5(2), 103, 1983.
20. National Academy of Sciences, *Risk Assessment in the Federal Government: Managing the Process*, National Academy Press, Washington, D.C., 1983.
21. National Academy of Sciences, *Risk Assessment in the Federal Government: Managing the Process*, National Academy Press, Washington, D.C., 1983, 21.
22. National Academy of Sciences, *Risk Assessment in the Federal Government: Managing the Process*, National Academy Press, Washington, D.C., 1983, 18.
23. Corn, J. and Corn, M., Changing approaches to assessment of environmental inhalation risk: a case study, *Milbank Quart.*, 73, 1, 1995.
24. Carson, R., *Silent Spring,* Houghton Mifflin Co., Boston, MA, 1962.
25. Murray, T., Regulating asbestos: ethics, politics, and scientific values, *Sci. Tech. Hum. Values,* 11(3), 13, 1986.
26. Ruckelshaus, W.D., Science, risk and public policy, *Science*, 221, 1027, 1983.
27. Lowrance, W., *Of Acceptable Risk,* William Kaufman, Inc., Los Altos, CA, 1976.
28a. IUD V API, 448 U.S. 607, 1980.
28b. Corn, J., *Response to Occupational Health Hazards*, Van Nostrand Reinhold, New York, 1992, 61.
28c. Mintz, B.W., *OSHA History, Law and Policy,* Washington Bureau of National Affairs, 1984.
29. National Academy of Sciences, *Risk Assessment in the Federal Government: Managing the Process*, National Academy Press, Washington, D.C., 1983.
30. Lowrance, W., *Of Acceptable Risk,* William Kaufman, Inc., Los Altos, CA, 1976, 2.
31. Lowrance, W., *Of Acceptable Risk,* William Kaufman, Inc., Los Altos, CA, 1976, 8.
32a. Corn, J.K., Vinyl chloride, setting a workplace standard: an historical perspective on assessing risk, *J. Publ. Health Policy,* 497, 1984.
32b. Bayer, R., Ed., *The Health and Safety of Workers: Case Studies in the Politics of Professional Responsibility,* Oxford University Press, New York, 1988.
33. *Federal Register 43.* 5918 (1978) OSHA Occupational Exposure to Benzene: Permanent Standard.

34. *Federal Register 43.* 5931, 5946 (1978) OSHA Occupational Exposure to Benzene: Permanent Standard.
35. IUD v API, 448 U.S. 607, 1980.
36. National Academy of Sciences, *Risk Assessment in the Federal Government: Managing the Process*, National Academy Press, Washington, D.C., 1983.
37. U.S. Congress, Office of Technology Assessment, *Researching Health Risks.* OTA-BB5-570, Washington, D.C., U.S. Gov't Printing Office, Nov. 1993.
38. General Accounting Office, *Asbestos in Schools: A Dilemma*, General Accounting Office, Washington, D.C., 1982, 2.
39. General Accounting Office, *Asbestos in Schools: A Dilemma*, General Accounting Office, Washington, D.C., 1982, 2.
40. *Federal Register,* Vol 47, #103, 23362, May 27, 1982.
41. *Federal Register,* Vol 47, #103, 23362, May 27, 1982.
42. *Federal Register*, Vol 47, #103, 23362, May 27, 1982.
43. *Federal Register*, Vol 47, #103, 23363, May 27, 1982.
44. Bernarde, M., *Asbestos: The Hazardous Fiber*, CRC Press, Boca Raton, FL, 1990, 277.
45. *Guidance for Controlling Friable Asbestos-Containing Material in Buildings*, U.S. EPA, OPTS, Washington, D.C., March, 1979.
46. Bernarde, M., *Asbestos: The Hazardous Fiber*, CRC Press, Boca Raton, FL, 1990, 279.
47. Bernarde, M., *Asbestos: The Hazardous Fiber*, CRC Press, Boca Raton, FL, 1990, 283.
48a. Corn, M., Airborne concentrations of asbestos-in-air in buildings and exposure of occupants: risk and regulatory implications, *Indoor Air,* 4, 491, 1991.
48b. Health Effects Institute/Asbestos Research, *Commercial Buildings: A Literature Review and Synthesis of Current Knowledge*, Cambridge, MA, HEI-AR, 1991.
49. Mossman, B.T. and Gee, J.B.L., Asbestos-related diseases, *N. Engl. J. Med.,* 320, 1721, 1989.
50. Mossman, B.T., Bignon, J., Corn, M., Seaton, A., and Gee, J.B.L., Asbestos: scientific developments and implications for public policy, *Science,* 247, 294, 1990.
51. Abelson, P.H., The asbestos removal fiasco, *Science*, 247, 4946, 1990.

3 The Making of an Environmental Issue

BEGINNINGS OF ASBESTOS IN SCHOOL POLICY

SETTING THE STAGE

This chapter focuses on the defining and emergence of the public health issue: asbestos in schools. It asks the question: How did asbestos become a major societal problem that would require large resources, cause both fear and consternation on the part of the public, and create controversy in the scientific community? It emphasizes the role that the United States Environmental Protection Agency (EPA), especially in the use of its guidance documents, and the media's reporting of health effects associated with asbestos had in forming public perception of asbestos and health, and shaping the issues that subsequently led to formulation of public health policy for asbestos in schools, a policy reflected in federal legislation, the Asbestos Hazard Emergency Act (AHERA).

Neither the media nor EPA placed risks from non-occupational exposure to asbestos into perspective. In the absence of public information that would place public health risks from asbestos in buildings in an accurate frame of reference, nonscientific factors, public perception informed by the message presented, and the past history of asbestos in the workplace inordinately influenced the establishment of public policy. A science-based issue became a value issue. Science, which should have been the deciding factor, took a back seat. Furthermore, the early presentation to the American public of knowledge about asbestos occurred simultaneously with rapidly rising concern about the environment, unfolding information about toxic substances, and fear of environmentally-caused cancer.

In 1970, the growing public interest in the environment found popular expression in Earth Day, celebrated by millions of Americans.[1] Congress and the Nixon Administration responded to the public call for action expressed by the media, Earth Day participants, consumer groups, and environmentalists when they passed a number of strict new environmental laws. In a short period of time the federal government had put into place an array of detailed regulations and policies to control environmental risks. Although many of the new environmental laws were passed during other administrations besides that of Nixon, it should be noted that both EPA and the Occupational Safety and Health Administration (OSHA) came into existence during the Nixon Presidency. The obstacles presented by the Nixon administration to carrying out the latter, newly-formed agencies' mission are yet another story.

The new legislation encompassed an enormous area of environmental concerns including air, water, pesticides, consumer protection, and occupational safety and health. It dealt with toxins as well as carcinogens. Congress passed many of the laws in response to the then current notion that the majority of cancers were environmental and occupational in origin. Table 3 in Chapter 2 lists major environmental legislation from 1938 to 1986.[2] The earlier laws covered air, water, and pesticides. In the 1970s, environmental policy shifted away from ecology, becoming more oriented to human health. Most environmental legislation was enacted between 1969 and 1980.

After World War II, the explosive growth in production of synthetic fibers, plastics, detergents, and pesticides created new environmental issues. Chemicals contaminated the air inside factories, as well as outside. Americans slowly became aware of air and water pollution. Although a movement for clean air and clean water had been developing along with other environmental activities, such as occupational hygiene, the American public was either unaware or chose to ignore pollution (a manifestation of environmental problems) except in a few rare instances when addressed by weak city or county regulations. Regulation of the environment was considered a responsibility of the states, and few states had effective environmental agencies or statutes.[3]

Occasionally something did occur to alert Americans and make them aware of the problems associated with pollution. An air pollution episode in Donora, Pennsylvania, was one of those early occasions. In 1948, a temperature inversion lasted for eight days in the town of Donora, and caused a severe air pollution episode. Twenty people died and hundreds were hospitalized as a result of the incident.[4] Donora aroused interest in air pollution, alerted the American public, and focused attention on the environment and upon the idea that chemicals from industrial sources can potentially and severely damage natural resources as well as human health.

In the 1950s, environmental worries and tensions focused on the radioactive debris that resulted from testing atmospheric nuclear weapons and fear of the effects radioactivity could have on the environment and human health. In the 1960s, furor over pesticides and other potential environmental contaminants surfaced when Rachel Carson's book *Silent Spring*[5] generated interest and concern. She protested the way pesticides were used and abused. Carson's message jolted Americans, got their attention, and stimulated their interest in the environment.

Other environmental episodes occurred in the 1950s and 1960s. It became easy to believe that air and water pollution were widespread and unchecked. In a special edition published by *Chemical and Engineering News* on EPA's twenty-fifth anniversary, it described the environmental situation in the United States on the eve of Earth Day, April 22, 1970.

> When EPA came into being, air and water pollution were rampant and obvious. Industrial smoke stacks and vehicular tailpipes belched out so much gunk that the fouled air actually killed the sick, the very young and the elderly. Rivers were so polluted that one, the Cuyahoga running through Cleveland actually caught on fire. Drinking water, the staff of life, was breeding life-harmful bacteria and other microscopic creatures — and collecting cancer-causing organic chemicals and brain damaging heavy metals such as lead."[6]

Although this 1995 description of pre-EPA conditions seems overdrawn, it does present an insight into public perception of environment in the 1960s. Environmental activists utilized this perception and helped create a public outcry for environmental regulation.

In response to demand for national environmental policy, and fully understanding the political implications, Richard Nixon established a special commission. The commission recommended that environmental programs, then scattered throughout the federal government, be consolidated into one agency. When Nixon created EPA, responsibility for pollution control became national. EPA could establish federal standards and program requirements for environmental protection, thus making pollution control national. EPA's goals included protection of environment and health for all Americans. The federal government had assumed an unprecedented amount of responsibility. Early mandates included:

1. The Clean Air Act of 1970 which required EPA to set ambient air quality standards, control mobile and stationary source emissions, and hazardous air pollutants;
2. Federal Environmental Pesticides Control Act of 1972 which amended the Federal Insecticide, Fungicide and Rodenticide Act and required manufacturers to supply toxicological information and register pesticides with the agency;
3. Banned D.D.T.;
4. The Federal Water Pollution Control Act Amendments to rewrite laws, expand sewage treatment, and establish goals of fishable, swimmable water.

EPA's early problems stemmed in part from the agency's inability to overcome the drag of tradition created by the method of forming the agency which brought together several predecessor programs and agencies. Programs, staffs, responsibilities, and institutional cultures clashed in the new agency and were often at odds with one another. The newly formed EPA encompassed programs previously administered by three departments, three bureaus, three agencies, two councils, one commissioner, one service, and several small offices.[7] For example, the Federal Water Quality Administration and all its pesticide programs came from the Department of the Interior. The National Air Pollution Control Administration, the Food and Drug Administration Pesticide Research, the Bureau of Solid Waste Management, the Bureau of Water Hygiene, and parts of the Bureau of Radiological Health came from the Department of Health Education and Welfare. The Agricultural Research Service of the Department of Agriculture contributed its pesticide activities. The Atomic Energy Commission and the Federal Radiation Council had their radiation criteria and standards programs placed in EPA. The Council on Environmental Quality transferred its ecological research to the new agency.

The first EPA administrator was William D. Ruckelshaus, a lawyer appointed by Richard Nixon and approved by Congress. He had a staff of 6,000 and a budget of $1.3 billion. Ruckelshaus left a legacy of standard setting, strong enforcement of anti-pollution laws (command and control), and the dominance of law over science.[8]

The purely political reasons dominating Nixon's organization of the EPA caused divisions within the agency resulting from bringing together a number of disparate agencies and offices. The consequent divisions within the agency and the pressure extended by the Congress and the public also account for the many problems EPA faced from its beginning. As a result, the agency responded to public pressure with instant solutions to complicated problems. Poorly-conceived and poorly-executed policy followed. Asbestos policy formulation at EPA fits this pattern.

EPA FOCUSES ON ASBESTOS

At the end of the 1970s, after concern that health hazards related to asbestos exposure had spread from occupationally exposed populations to those experiencing non-occupational exposure, exposures of school children became the focus for the issue. Anxious Americans reacted in a panic to the fact that building materials which had been utilized in school construction for 30 years or longer contained asbestos. Why did the American public respond in such a manner? Why did the asbestos policy evolve from an earlier one that required asbestos in building materials (for safety and protection from fire) to a later policy of removal, reflected in the AHERA of 1986?

In 1989, EPA estimated that 733,000 United States public and commercial buildings contained friable,* asbestos-containing materials (ACM), and that between 160,000 and 250,000 abatement jobs were performed yearly on all types of buildings, with abatement activity in schools much higher than other public buildings. 28% of schools had full removal.[9] Billions of dollars were spent to remove asbestos from buildings, based on a public policy decision that was scarcely questioned** until 1990.

EPA regulated ACM with two goals. The first, from the standpoint of air pollution, was intended to prevent the release of fugitive asbestos fibers into the air outside of all buildings. EPA added asbestos to the list of regulated materials in 1971 under the National Emission Standard for Hazardous Air Pollutants (NESHAP). EPA revised the NESHAP in 1973 to set a qualitative standard of no visible emissions for milling, manufacture, and demolition of ACM. At that time, EPA could only regulate to prevent emission of asbestos to the environment outside buildings; OSHA retained jurisdiction inside buildings when commercial users of asbestos were involved. EPA's major asset to regulate asbestos was the NESHAP. The second goal of EPA concerning asbestos was to protect a segment of the population namely, school children, from adverse health effects of airborne asbestos inside buildings. The initiative to protect school children originated in the EPA Office of Pesticides and Toxic Substances (OPTS). EPA rationalized the need to protect school children in the following manner. EPA believed the risks of adverse health effects grew with increased dose, either in concentration of asbestos or over time of exposure. Thus, a person exposed for a longer period of his or her lifetime had a greater risk of developing asbestos-related disease. A child exposed to asbestos in schools had a greater lifetime risk of developing mesothelioma, for example, than did a janitor

---

* Friable = crumblable by hand.

** The term *scarcely* is not meant to overlook the few scientists who questioned asbestos in school policy, but were largely ignored until 1990.

exposed for the first time at age 50.[10] William Nicholson, a professor at the Mt. Sinai School of Medicine, Environmental Sciences Laboratory, articulated this concept in numerous documents prepared under contract for EPA. For example, in the document *Airborne Asbestos Health Assessment* Nicholson concluded: "Children exposed at younger ages are especially susceptible because of their long life expectancy. The time of exposure plays little role in the lifetime excess risk of lung cancer; any exposure before the age of 45 or 50 contributes equally to the lifetime risk."[11]

In 1979, given EPA's acceptance of the theory that asbestos in school buildings created an unacceptable risk to children and the agency's legal mandate, the Toxic Substances Control Act (TSCA), EPA instituted its Technical Assistance Program (TAP). TAP was a nonregulatory attempt to educate school officials, state government officials, parents, and school employees about the hazards of airborne asbestos. TAP sought to educate school officials about asbestos hazards and to encourage voluntary identification and correction of problems created by friable asbestos in school buildings. In 1982, EPA promulgated the Asbestos-In-Schools-Identification and Notification Rule, which required inspection for friable asbestos in all schools and notification to parents, teachers, and school workers.[12] Congress passed the Asbestos School Hazard Abatement Act (ASHAA) in 1984 to provide financial assistance to needy schools with serious asbestos hazards.[13] On October 22, 1986, President Ronald Reagan authorized AHERA. Representative James Florio of New Jersey had won unanimous passage of this legislation, which he proposed and shepherded through the Congress. AHERA required all public and private schools to inspect for asbestos, develop asbestos management plans, submit them to the state, and implement appropriate response actions and abatement.

In the case of asbestos policy, the political origins and orientation of the agency offer some insight into the asbestos program. The program was enveloped in both misunderstanding about the role of the agency and mistaken scientific premises. There was little debate within EPA about the nature of the issues, the relationship of science to environmental problems, or even about science and policy. Discussions of strategies about who should be protected, how to protect, and the cost of protection cannot be found in agency documents. More public debate spearheaded by the agency could have better framed questions and issues. Public debate and education might even have taken into account the limits of available science, engineering, and managerial understanding.[14] At the same time the Congress drove EPA by defining the agency's authority and responsibility while allocating needed funds. But the political origins and shortcomings of EPA alone cannot fully account for why the American people apparently accepted asbestos policy for schools without question, and the fiasco that followed.

Other factors helped to define the asbestos in schools issue and to shape public perception which subsequently led to the formulation of public health policy for asbestos in schools, a policy reflected in major federal legislation, AHERA. Past history of asbestos in the workplace (including the legacy of death and painful illness from shipbuilding during World War II), the asbestos awareness campaign, rapidly rising public concern about environmentally caused cancer, and media presentation of information about toxic substances, including asbestos, all help to explain and set the stage for Americans' response to asbestos in schools.

LEGACY OF ASBESTOS IN THE WORKPLACE

The public health consequences of past, poorly-controlled asbestos in the workplace can be traced to decisions made over 50 years ago during World War II when shipyards and other defense industries extensively utilized asbestos with minimum controls and minor precautions, leaving a legacy of disease and death. Manifestations of asbestos disease appeared with increasing frequency 20 years after World War II, affecting thousands of shipyard and other workers who had been exposed in the 1940s and during the postwar period when injudicious applications multiplied the amount of asbestos used.[15] In 1981, Peto et al. estimated that exposure prior to 1965 would cause approximately 37,500 mesothelioma deaths and 112,500 lung cancer deaths.[16] Nicholson and others estimated that past exposures to asbestos would cause from 8,700 excess deaths from cancer per year in 1982 to almost 10,000 per year by 1990. After 1990, about 9,000 such deaths per year would occur until the 21st century when the number would begin to decline.[17] Doll and Peto estimated that in 1978 perhaps 5% of the lung cancers in the United States could be attributed to occupational exposure to asbestos, an estimated 4,000 to 8,000 deaths per year.[18] The work of Selikoff et al., citing the occurrence of asbestosis and lung cancer in insulation workers, published in the 1970s had the greatest impact upon EPA decisions about asbestos. EPA utilized it in risk assessment documents, the Federal Register, and support documents. The Selikoff insulation workers' study documented the mortality experiences of 17,800 insulation workers who had been enrolled in the International Association of Heat and Frost insulators and Asbestos Workers in 1967. Selikoff et al. observed 2,271 deaths between 1967 and 1976. The research made it clear that the risk of asbestos disease was not confined solely to workers in mining and manufacturing. It extended to those who used asbestos products, for example, shipyard workers, insulation workers, and many others outside primary or fixed place industries.[19]

As noted earlier in Chapter 1, it took many years to proceed from initial knowledge to confirmation and general acceptance of asbestos as a cause of cancer. American society, as distinct from a relatively small core of public health and technical professionals, had not been widely sensitized to the idea that yesterday's environment could contribute to today's illness. In 1970, enactment of the watershed legislation that created OSHA alerted the general public about occupational health and the possibility that past employment could contribute to future disease. Industrial response to occupational risks associated with asbestos shaped attitudes and perception about non-occupational risks. The legacy of distrust resulting from past neglect of workplace exposures, manifested in the proliferation of lawsuits, informed public debate about non-occupational exposure.

Litigation of personal injury claims occurred as asbestos workers of the 1940s and 1950s came forward with large numbers of asbestosis, lung cancer, and mesothelioma cases. Because personal injury claims precipitated by occupational exposures were highly publicized, Americans learned that the legacy of asbestos in the workplace was chronic lung disease. Tens of thousands of workers filed suits against asbestos manufacturers claiming asbestos made them ill with asbestosis, cancer, or mesothelioma. Asbestos companies spent millions of dollars to defend or settle claims. The lawsuits pointed out negligence and misrepresentation on the part of the asbestos industry.

The growing realization of the risks associated with high levels of asbestos in the workplace, even though many occurred in the past, and the short sightedness of many asbestos companies in the early years of the 20th century, aroused deep feelings about asbestos. Anger about past injustices welled up. Apparently many men in responsible positions within companies knew about the risks associated with high levels of asbestos at work, did not accept the validity of early scientific research, and did little or nothing to protect or compensate workers. Much of this information unfolded during discovery and in the courtroom.[20] It was difficult not to sympathize with workers who were often unaware of danger and became ill because of work with asbestos. A number of books and journal articles highly critical of the role played by the asbestos industry were published. For example, Paul Brodeur's book, *Expendable Americans*, first published as a series of articles in the *New Yorker*, informed a select readership about the issue of asbestos and occupational disease.[21] In *Outrageous Misconduct*, also first published in the *New Yorker*, Brodeur sung the praises of plaintiffs' lawyers whose clients sued the asbestos companies for large sums of money.[22] Most of the heavy workplace exposures occurred prior to the Occupational Safety and Health Act of 1970 and the Asbestos Standard the agency promulgated in 1971.

Information was not presented to the public concerning the difference between high and low dose, or between public buildings and the workplace. People were fearful and uncertain. Furthermore, because breathing asbestos can cause lung cancer the widely held "one fiber can kill" theory did not help to further understanding or acceptance of the idea that asbestos in low dose is not as dangerous as in high dose. The difference between high and low dose and between occupational and non-occupational exposure to asbestos was never properly expressed. The difference was blurred and the historical experience of workers in the 1930s, 40s, 50s, and 60s only led to concern about asbestos, no matter where it was found or what the dose or time of exposure. It made Americans fearful that their children might be at risk in school.

ASBESTOS AWARENESS CAMPAIGN

In 1978, a message about the hazards of working with asbestos was sent to the American people by the federal government in the form of the "Asbestos Awareness Campaign." Based on epidemiological studies of Irving Selikoff and others in the 1960s, which established that certain groups suffered unusually high rates of cancers, including mesothelioma, a disease virtually unknown before the 20th Century,[23] the campaign was an effort to inform former and current workers and others at risk about the hazards associated with asbestos exposure. The objective was to increase awareness and to present information about the nature, extent, and seriousness of asbestos exposure for the at-risk target audience and the general public. The campaign disseminated information about asbestos over radio and television, by print, through relevant special interest groups, and via direct mail.

A study of the "Asbestos Awareness Campaign" noted that difficulty influencing the media could be traced to "gatekeepers," for example, editors who controlled the flow of the news. The study also revealed that nearly 90% of the asbestos news was accounted for by three topics: (1) hazards of asbestos exposure (50%); (2) negligence or cover-ups (23%); and (3) governmental action to control or reduce risk (15%).[24]

ENVIRONMENT AND CANCER

In the 1970s, the focus on environment shifted from an earlier concern with the impact of human activities on the environment and obvious pollution problems to a newer, less visible and difficult concern with human health, manifested in the "War on Cancer." Fear of cancer and the fostering and formalization of the idea that any amount of a carcinogen caused cancer was the rationale for cancer policy. For example, EPA cancer policy was based on the premise that a threshold for carcinogens (i.e., asbestos) did not exist. The rallying cry was "one fiber can kill." This occurred during a period of intense environmental activity and new legislation. Claims that an inordinately large number of cancers were caused by exposure to industrial toxins gained popularity. The belief that most cancers were environmentally caused, and thus preventable, influenced the thinking of most environmentalists, and informed public policy. It followed that if environmental factors are important the war on cancer should seek to prevent environmental contamination, both occupational and non-occupational, rather than place emphasis on the search for the magic bullet, i.e., a cure for cancer. In his book *Cancer Wars*, Robert Proctor wrote, "... though scholars would use the term "environmental factors" broadly, leading to confusion as to how one should go about the business of prevention. It was never easy to determine, for example, exactly what proportion of cancers were attributable to industrial effluents and what proportion were to be assigned to other environmental factors such as smoking, alcohol, dietary fat, and viral infection."[25] Environmental activists such as Samuel Epstein clearly blamed industry as a source of cancer from chemical carcinogens. He convinced many that most human cancers were caused by chemical carcinogens in the environment. Epstein accused industry of suppressing and destroying cancer data and regulatory agencies of being subverted by political pressure.[26] Critics of Epstein said he exaggerated the magnitude of environmental hazards, and placed too much emphasis on popular action to prevent environmental cancer. Richard Peto, a well-known epidemiologist, suggested that cancer research was in danger of becoming polarized because environmental zeal such as Epstein's distorted scientific judgment.[27] According to other scientists, man-made chemicals contributed relatively little to the total cancer burden.[28]

Another factor that influenced public concern about cancer and the environment, commonly referred to as the Estimates Paper, was a draft report entitled, "Estimates of the Fraction of Cancer in the United States Related to Occupational Factors."[29] The Secretary of Health Education and Welfare (now Health and Human Services), Joseph Califano, told an AFL-CIO conference on occupational safety and health that at least 20% of all cancers in the United States may be work-related and that 10 to 15% will be associated with previous exposure to asbestos. It is now understood that the estimates were grossly overstated. Nevertheless, it helps to explain popular sentiment about occupationally-induced cancer.

Some comments from the scientific community stated a belief that the authors of the Estimates Paper overstated the proportion of cancer traceable to the workplace.

> I regard it (the Estimates Paper) as scientific nonsense.[30]

> It shows how a group of reasonable men can collectively generate an unreasonable report.[31]

> It seems likely that who ever wrote (the Estimates Paper) ... did so for political rather than scientific purposes, and it will undoubtedly continue into the future as in the past to be used for political purposes by those who wish to emphasize the importance of occupational factors.[32]

> This report immediately received wide publicity; it was quoted by the Secretary of Health Education and Welfare and obtained coverage in the world news. Yet its frame is unsubstantial ... it is sad to see such a fragile report under such distinguished names.[33]

On the positive side, concern about cancer did provide impetus for preventive efforts to reduce its incidence. Spurred by the association between environment and cancer, even if overdrawn, public and congressional interest led to enactment of legislation to provide federal agencies with mechanisms for regulatory intervention. Table 1 indicates public laws providing for the regulation of exposures to carcinogens. It lists the legislation (agency) and agents regulated as carcinogens. Information for the table is from an Office of Technology Assessment document, *Assessment of Technologies for Determining Cancer Risks from the Environment*.[34]

RESPONSE TO REACTIONARY REAGAN YEARS

In the 1980s, reaction to Reagan environmental policy embodied in the shift away from enforcing regulations, corruption at Superfund, dismantling of environmental statutes, disarray at EPA and the resignation of Mrs. Gorsuch, EPA administrator, engendered a feeling of both anger and frustration, and a need for environmentalists to regroup around an issue that would arouse Americans. Asbestos in schools was that issue. The Environmental Defense Fund, for example, played a leading role. Unlike asbestos in the workplace, non-occupational asbestos could affect everybody, including the most helpless Americans: children. It was an involuntary risk. It is not surprising that Congress unanimously passed AHERA as the culmination of asbestos in schools policy during the Reagan administration and that the president signed it.

INFORMING THE PUBLIC

It is impossible to focus on the defining and emergence of a public health issue — asbestos in schools — without emphasizing the role played by media reporting of health effects associated with asbestos, and EPA. The media and EPA defined the issue and shaped public perception that subsequently led to creation of public health policy for asbestos in schools, a policy reflected in major federal legislation,

TABLE 1
Public Laws Providing for the Regulation of Exposure to Carcinogens (1981)

Legislation (Agency)	Agents Regulated as Carcinogens or Proposed for Regulation
Federal Food, Drug and Cosmetic Act (FDA)	
food	21 food additives and colors aflatoxins — PCBs — nitrosamines
drugs	not determined
cosmetics	not determined
Occupational Safety and Health Act	20 substances
Clean Air Act	
Sec.112 stationary sources	Asbestos, beryllium, mercury, vinyl chloride, benzene, radionuclides, and arsenic (an additional 24 being considered)
Sec. 202 vehicles	Diesel particulate standard
Sec. 211 fuel additives	
Clean Water Act (EPA Sec. 307)	49 substances listed as carcinogens by CAG
Federal Insecticide, Fungicide and Rodenticide	14 rebuttable presentations against registrations either initiated or completed: nine pesticides voluntarily withdrawn from the market
Resource Conservation and Recovery Act (EPA)	74 substances proposed for listing as hazardous wastes
Safe Drinking Water Act (EPA)	Trihalomethanes, chemicals formed by reactions between chlorine, used as disinfectant and organic chemicals. Two pesticides and two metals classified as carcinogens by CAG but regulated because of other toxicities.
Toxic Substances Control Act (EPA)	
Sec. 4 (to require testing)	six chemicals used to make plastics pliable
Sec. 6 (to regulate)	PCBs regulated as directed by law
Sec. 7 (commence civilization against imminent hazards)	
Federal Hazardous Substances Act (CPSA)	
Consumer Product Safety Act (CPSA)	five substances: asbestos, benzene, benzidine (and benzidine-based dyes and pigments), vinyl chloride, "Tris"

AHERA. Neither the media nor EPA placed risks from non-occupational exposure to asbestos into perspective. In the absence of information that would place public health risk from asbestos in buildings in a reasonably accurate perspective, nonscientific factors, public perception informed by the knowledge presented, and the past history of asbestos inordinately influenced the establishment of the public health policy. The subject of perception of risk is a broad one and the current literature which utilizes sociological, physiological, and political science concepts has been helpful. It points out that concerns about risk are based largely on qualitative rather than quantitative factors.[35]

The Media and Public Information

Looking backward, it is clear that presentation to the public of health information about asbestos coincided with rapidly rising concern about the environment, issues of toxic substances, fear of cancer, the legacy of occupational asbestos, and the response to the reactionary Reagan administration. Placed in that social context, the manner and content of information presented undoubtedly influenced public perception of the asbestos hazard in schools. The most likely source of information for most people — the media — amplified the risks as it framed the discussion. In the process, they reported on harm from earlier high levels of asbestos based on industrial experience and not risks in schools which were orders of magnitude lower than in industry. The media tended to accept information from dominant institutions involved in this issue and failed to place health risks into perspective. The media failed to emphasize or discuss issues such as the likelihood of exposure to asbestos in the school environment, the ratio of benefits to risks of removal, the cost of removal, or alternative options. Asbestos in schools was served up as a crisis.

Although journalists did seek individuals with authoritative knowledge, the sources they utilized decided what material to reveal, what details to highlight or discard, and followed their own agendas. In the period under discussion, the main sources of information at the disposal of journalists on the subject of asbestos came from Dr. Irving Selikoff and his colleagues at the Mount Sinai School of Medicine, Environmental Science Laboratory, in New York and EPA. Their view was that asbestos, a tragedy for industrial workers, also posed a great threat to the general public. This viewpoint appeared in countless newspaper articles as well as other media. The message said evidence warranted removal of asbestos already in place in schools because asbestos posed an unacceptable and intolerable health threat to 15 million American school children. The following representative messages appeared during this period.

- *New York Times*, July 18, 1971: "Asbestos You'll Be All Right If You Just Stop Breathing." Discussion of a number of aspects of environmental asbestos with a message that asbestos is everywhere. The metaphor of the "asbestos time bomb" appeared and would appear again and again. Dr. Selikoff was the author's information source.[36]
- *New York Times*, September 30, 1972, author Jane Brody: "Shipyard Workers of 1940s Told of Cancer Peril." The article is about Dr. Selikoff's

epidemiological studies of shipyard workers. "Now the doctor said, it appears that even people who did not work directly with asbestos, but who were indirectly exposed to asbestos particles may also be in danger of developing the disease which tended to occur about 30 years after the initial exposure to asbestos."[37]

- *New York Times*, September 22, 1974: "The Contagion of Asbestos." It said, "Mount Sinai has produced evidence to show that even peripheral and brief exposure can produce asbestos-related disease."[38]
- *New York Times*, November 21, 1974: "Pupils Move Today To Avoid Asbestos." Article said, "That when released into the air from construction materials, asbestos is believed to cause cancer and respiratory disease."[39]
- *Wall Street Journal*, November 3, 1978: "EPA Set To Require Asbestos Inspections By School Districts." Article said, "That the Environmental Defense Fund petitioned EPA earlier to order school districts to require asbestos inspection. The EDF contended that 20 million school children may be exposed to risk of asbestos-induced cancer or lung disease."[40]
- *New York Times*, March 31, 1980: "Penn Yan Schools to Act on Asbestos." Article said, "All schools in Penn Yan Central School District will be closed today and tomorrow because of the discovery of "significant amounts" of asbestos particles in the building."[41]
- *Wall Street Journal*, February 23, 1984: "U.S. Considering Rules to Clean Up Asbestos in Schools." A recent internal EPA report concluded that hundreds of schools with serious asbestos pollution haven't done anything to alleviate the danger. The Service Employees International Union and the National Education Association asked for more stringent regulations.[42]
- *U.S.A. Today*, February 2, 1984: "School Asbestos Clean Up Lagging." Less than a fifth of the 2,632 asbestos-plagued schools in the U.S. have plans to correct the problem. The article said that fibers are crumbling and that inhaled asbestos fibers have been found to cause cancer and other lung problems.[43]
- *New York Times*, May 8, 1984: "EPA is Pressed to Require Removal of Asbestos." Scientists, union officials, and parents urged EPA to require the removal of the carcinogen asbestos from schools. Irving Selikoff is quoted as saying, "Ultimately what you will do is decide who will live and who will die."[44]

The television media caused alarm and misunderstanding of the issues. The following are two examples taken from transcripts of television reports in 1979. On CBS, Bob Schieffer introduced an interview with Douglas Costle, EPA Administrator. The discussion was centered on an EPA guidance document. Schieffer said: "Asbestos has been considered a cancer threat for some time, and now the federal government is urging quick action by local authorities to eliminate hazards."[45] Although Costle tried to keep the interview sensible, a tone was set by the introduction.

On ABC, Max Robinson introduced the subject of asbestos in schools. "The federal government today called for a nationwide campaign to protect children from cancer-causing asbestos in school buildings. Herbert Kaplow has a report." Kaplow:

"It may be on the walls and ceilings of as many as ten thousand of the nation's ninety thousand public schools. And the Environmental Protection Agency says that could be dangerous, because asbestos tends to flake, and the particles can be inhaled by children and twenty years later they could find themselves with lung cancer or other serious illness."[46] What's a mother to think?

The news commentators had so little understanding of the issue they were discussing that they presented inaccurate information and pressed interviewees to quickly answer complicated questions in one sentence. This could only lead to misunderstanding. And it did.

EPA GUIDANCE

In 1979, when EPA embarked on its public policy for asbestos in schools by assisting schools in "identifying and controlling asbestos-containing materials in their buildings," it instituted the Technical Assistance Program (TAP). The goal of TAP was to encourage voluntary identification and correction of problems caused by asbestos materials.[47] But EPA, in its goal to protect the public, presented a position to the public which was scientifically and technically insupportable. EPA played a major role in the setting and carrying out of public policy for asbestos in schools.

During policy development, EPA ignored the scientific and technical paradigm developed by industrial hygienists, namely that of recognition, evaluation, and control of potential hazards.[48] During EPA policy development the concept of evaluation or exposure assessment, a paradigm ingredient which necessitates measurement of exposure, was bypassed. EPA guidance documents for control of asbestos in schools apparently neglected the already developed scientific underpinnings necessary for sound environmental policy. For example, EPA did not endorse air sampling to measure airborne concentrations of asbestos in schools. Industrial hygiene practice commonly employs air monitoring to evaluate exposures to a wide range of airborne contaminants. EPA also failed to identify a safe or acceptable level of risk and adopt guidelines governing non-occupational exposure to asbestos, an activity synonymous with the practice of industrial hygiene.

TLV (threshold limit value) for limiting exposure to a toxic material, a bedrock concept of industrial hygiene since 1947, is associated with airborne concentrations of potentially toxic materials measured with instruments.[49] A series of toxicological and epidemiological studies were also utilized by industrial hygienists to relate what people breathed to a PEL (permissible exposure limit). All of this preceded concern for asbestos in the environment and was already in place by 1971 when OSHA opened its doors. Furthermore, EPA had set ambient air quality standards, based on industrial hygiene methods, for allowable concentrations in the community by the time the asbestos-in-schools issues arose at EPA. Therefore, it was an accepted and well-established procedure that if a potential inhalation risk existed, you measured the concentration in air to evaluate the risk. EPA scientists and policy makers ignored this approach.

Since 1979 EPA has published a series of state-of-the-art technical guidance documents which have been widely distributed to schools, states, architects, and other professionals involved in identification and control of asbestos. The guideline documents are referred to as *The Orange Book* (1979),[50] *The Blue Book* (1983),[51] *The Purple Book* (1985),[52] and *The Green Book* (1990),[53] based on the color of their covers.

Early on, *The Orange Book* crystallized EPA's view of the role of air sampling in evaluating the risk of exposure of building occupants to airborne asbestos. The purpose of the document was to act as a guide for deciding if asbestos posed a problem. *The Orange Book* sought to get people to make environmental decisions without utilizing already established guidelines such as TLVs or PELs; It asked the question, "Is air sampling necessary to evaluate exposure potential?" It was answered in the following manner: "Air sampling is inappropriate to estimate asbestos contamination and exposure. In the school environment, it is virtually impossible to establish exposure potential using standard air sampling technique."[54] This violation of the principle of air sampling initiated asbestos policy for schools and public buildings. It set a tone that ignored risk assessment.

Instead of relying on the well-established principle of air sampling, i.e., quantitative risk assessment, EPA adopted an algorithm, a device that can be viewed as a tool for qualitative risk assessment. The algorithm consisted of a series of eight environmental factors: condition of material, water damage, exposed surface area, accessibility, activity and movement, air plenum or direct airstream, friability, and asbestos content.[55] Numerical weighting factors were assigned to each environmental factor and a formula used to derive a numerical score for the given situation. Based on the algorithm score, the responsible party was directed to take immediate action, i.e., removal, or to postpone action. In most cases the algorithm was biased toward immediate action. In 1979, neither the professional community nor the public knew that the algorithm in the first guidance document had never been tested for its correlation with airborne asbestos concentrations in different environments. Subsequently, two studies contracted by EPA indicated the poor correlation of the algorithm scores with airborne fiber measurements.[56] Water damage to the material was the only consistent environmental factor. Other investigations of EPA algorithm reported similar conclusions.[57]

The need to correlate an algorithm with airborne fiber concentrations is necessary if risk is to be estimated by a method other than measuring airborne concentrations. The presence of a source or an agent is a potential hazard, but the agent must be transmitted from the source to the human receptor and breathed if a potential hazard is to become a real hazard. In-place asbestos in a building is a potential hazard; airborne asbestos is a real hazard. The algorithm equated a potential hazard with a real hazard without validating the equivalency. In fact, the equivalency did not exist.

In 1982, a rule entitled, "Friable Asbestos-Containing Materials In Schools: Identification and Notification" required schools to inspect and take samples of materials to determine if asbestos was present, and to inform parents and workers in schools if it was.[58] Local educational agencies were required to keep a copy of the *The Orange Book* on file.

In 1983, *The Blue Book* discussed air sampling and the algorithm. On the subject of measuring airborne asbestos in a non-occupational environment, *The Blue Book* stated, "Another proposed approach to assessing the need for corrective action is to measure asbestos fibers in the air. At best this approach provides information only on current asbestos contamination and no information about the potential for fiber release and future air levels. Moreover, the use of air monitoring as an assessment tool involves substantial technical and economic problems which limit its use even for determining current levels of contamination ... Given the limitations, EPA does not recommend the use of air monitoring for assessment purposes at this time."[59]

In 1985, *The Purple Book* did not recommend air monitoring as a primary assessment tool at that time.[60] All three of the guidance books, covering a period of six years, elaborated on removal, operations, and maintenance methods. All three stressed removal and none acknowledged the relevance of industrial hygiene exposure assessment involving measurement of asbestos-in-air to policy for non-occupational exposure to asbestos in buildings.

AHERA, signed by Ronald Reagan in 1986, required over 100,000 American schools to conduct inspections to identify both friable and nonfriable asbestos-containing building materials, and based on the inspections to prepare a plan to manage asbestos safely.[61] AHERA reflected policy consistent with the three EPA asbestos guidance documents discussed.

The current messy situation with regard to asbestos risk in non-occupational settings results from the fact that EPA guidance documents never addressed the magnitude of the risks involved, and failed to relate asbestos risk to other environmental risks. Furthermore, the lack of EPA efforts to obtain rigorous measurement of asbestos fibers went unquestioned by most environmental scientists. One would think that the first priority should be accurate assessment of the problem. Failure to utilize the framework developed over a 50 year period by industrial hygienists had led to a costly policy. In the end, the cause of prudent environmental policy was ill served by action based on public perception driven by the dramatic and sensational, rather than proven or sound scientific and technical tenets.

In 1990, EPA published a fourth guidance document, *The Green Book,* acknowledging the relevance of industrial hygiene exposure assessment to policy for non-occupational exposure to asbestos in buildings.[62] A number of concerns had finally surfaced including the expense of removal of asbestos, the damaging aspects that come from removal, the stirring up of asbestos, and what to do with it once it is removed. Concern developed over whether all types of asbestos are equally hazardous, the lack of standards for exposure to asbestos, and the virtually nonexistent risk assessment. How reevaluation of policy for asbestos came about is the subject of a later chapter.

The first three guidance documents, as well as numerous press releases and reports, led to misconceptions, anxiety, and unnecessary removals. They fixated the idea of cancer and the one fiber theory in the public mind in relation to asbestos in schools and caused fear among parents and school administrators. EPA estimated premature deaths among school children in 1980 at 100 to 8,000 in 1981, from 40 to 400 in 1981, and in 1982 estimates were no longer stated.

Guidance documents contained inconsistencies in the agency's approach, ignored industrial hygiene practices, such as air sampling, to measure airborne concentrations of asbestos in schools, did not identify "safe" levels or acceptable levels of asbestos-in-air, and recommended removal of asbestos in schools. All three guidance books, covering a period of six years, stressed removal of asbestos. None acknowledged the relevance of exposure assessment to policy for non-occupational exposure to asbestos in schools. The guidance documents never addressed the magnitude of the risks involved or related asbestos risks to other environmental risks.

The role of EPA should have been to educate the American public. The agency seemed to prefer instant and simple solutions no matter how complicated or complex the problem. Critical questions about who should be protected, how to protect, why to protect, and the cost of protection were not publicly debated. Perhaps more public education and debate might have framed a more workable issue.

REFERENCES

1a. Hays, S., *Beauty Health and Permanence*, Cambridge University Press, 1987.
1b. Gottlieb, R., *Forcing the Spring*, Island Press, 1993.
2. See Chapter Two, 23.
3. Corn, J., *Protecting the Health of Workers: The American Conference of Governmental Industrial Hygienists*, ACGIH, Cinn. 1989, 69.
4. Corn, J., *Protecting the Health of Workers: The American Conference of Governmental Industrial Hygienists*, ACGIH, Cinn. 1989, 70.
5. Carson, R., *Silent Spring*, Houghton Mifflin Co., Boston, 1962.
6. Ember, L.R., View from the top: EPA administrators deem agency's first 25 years bumpy but successful, *Chemi. Eng. News*, 18, 1995.
7. Ember, L.R., View from the top: EPA administrators deem agency's first 25 years bumpy but successful, *Chemi. Eng. News*, 18, 1995.
8. EPA History Program, *William D. Ruckelshaus Oral History Interview, Document,* 93-202-K-003, Washington, D.C., 1993.
9. USEPA, *EPA study of asbestos-containing materials in public buildings: a report to Congress*, Washington, D.C., 560/5-88-0222, 1988.
10. USEPA, *Airborne asbestos health assessment update*, EPA/600/8-84/003F, Washington, D.C., 173, 1986.
11. USEPA, *Airborne asbestos health assessment update*, EPA/600/8-84/003F, Washington, D.C., 166, 1986.
12. *Federal Register, Friable asbestos-containing materials in schools: identification and notification*, EPA, May 27, 1982.
13. ASHAA, Public Law 98-377, 20 U.S.C. 4011, August 11, 1984.
14. Landy, M., Roberts, M., and Thomas, S., *The Environmental Protection Agency: Asking the Wrong Questions*, Oxford University Press, New York, 1990.
15. Corn, J. and Starr, J., Historical perspective on asbestos: policies and protective measures in World War II shipbuilding, *Am. J. Ind. Med.,* 11, 359, 1987.
16. Peto, J., Henderson, B.E., and Pike, M.C., Trends in mesothelioma incidence in the United States and the forecast epidemic due to asbestos exposure during World War II, in *Banbury Report 9: Quantification of Occupational Cancer*, Peto, R. and Schneiderman, M., Eds., Cold Spring Laboratory, Cold Spring Harbor, 1981, 51.
17. Nicholson, W. et al., *Banbury Report 9: Quantification of Occupational Cancer*, Peto, R. and Schneiderman, M., Eds., Cold Spring Laboratory, Cold Spring Harbor, 1981, 87.
18. Doll, R. and Peto, R., The causes of cancer: quantitative estimates of avoidable risks of cancer in the United States today, *J. Natl. Cancer Inst.,* 66, 1191, 1981.

19. Selikoff, J., Hammond, E.C., and Seidman, H., Mortality experience of insulation workers in the United States and Canada, *Ann. N.Y. Acad. Sci.*, 330, 91, 1979.
20. Castleman, B., *Asbestos Medical and Legal Aspects*, Harcourt Brace Jovanovich, New York, 1984, 401.
21. Brodeur, P., *Expendable Americans*, The Viking Press, New York, 1973.
22. Brodeur, P., *Outrageous Misconduct: The Asbestos Industry on Trial*, Pantheon Books, New York, 1985.
23. Selikoff, I.J. and Churg, J., Biological effects of asbestos, *Ann. N.Y. Acad. Sci.*, 132, 1965.
24. Freimuth, V. and VanNevel, P., Reaching the public: the asbestos awareness campaign, *J. Commun.*, 31(2), 155, 1981.
25. Proctor, R.N., *Cancer Wars*, Basic Books, New York, 1995, 55.
26. Epstein, S., *The Politics of Cancer*, Anchor Press, New York, 1979.
27. Peto, R., Distorting the epidemiology of cancer: the need for a more balanced overview, *Nature*, 284, 297, 1980.
28. Whelan, E.M., *Toxic Terror: The Truth Behind Cancer Scares*, Promethius Books, New York, 1993, 41.
29. NCI, NIEHS, and NIOSH, Estimates of the Fraction of Cancer in the United States Related to Occupational Factors, Sept. 15, 1978.
30. Doll, R., in American Council on Science and Health. ACSH *News and Views*, Nov.–Dec., 1979, p. 6.
31. Peto, R., *Nature*, 284, 1980.
32. Doll, R. and Peto, R., *The Causes of Cancer*, Oxford University Press, New York, 1981.
33. Whalen, E.M. Op. Cit. p. 50. Whalen quotes *Lancet* but does not site the journal issue.
34. Office of Technology Assessment, *Assessment of Technologies for Determining Cancer Risks from the Environment*, Congress of the United States, Washington, D.C., 178, 1981.
35a. Nelkin, D., *The Language of Risk*, Sage Publications, 1985.
35b. Gusfield, J.R., *The Culture of Public Problems*, University of Chicago Press, 1981.
35c. Rushefsky, M., *Making Cancer Policy*, SUNY Press, 1986.
35d. Slovic, P., Perception of risk, *Science*, 236, 1987.
35e. Singer, E. and Endreny, P., *Reporting on Risk*, Russell Sage Foundation, New York, 1984.
36. Asbestos: you'll be alright if you just stop breathing, *New York Times*, p. 6, July 18, 1971.
37. Shipyard workers of 1940s told of cancer peril, *New York Times*, p. 62, Sept. 30, 1972.
38. The contagion of asbestos, *New York Times*, Sept. 22, 1974.
39. Pupils move today to avoid asbestos, *New York Times*, p. 2, Nov. 21, 1978.
40. EPA set to require asbestos inspections by school districts, *Wall Street Journal*, p. 3, Nov. 3, 1978.
41. Penn yan schools to act on asbestos, *New York Times*, March 30, 1980.
42. Pasztor, A., U.S. is considering rules to clean up asbestos in schools, *Wall Street Journal*, Feb. 23, 1984.
43. Stevens, C., School asbestos cleanups lagging, *U.S.A. Today*, Feb. 2, 1984.
44. EPA is pressed to require removal of school asbestos, *New York Times*, May 8, 1984.
45. Transcript of CBS Station WDVM TV, *Friday Morning*, from OPTS docket #61004 #61-792-0235, March 16, 1979.
46. Transcript of ABC Station WJLA TV, *World News Tonight*, from OPTS docket #61004 #61-792-0239, March 16, 1979.

47. *Federal Register*, vol. 52, No. 83 April 30, 1987.
48. National Institute for Occupational Safety and Health, *The Industrial Environment: Its Evaluation and Control*, U.S. Government Printing Office, Washington, D.C., 1974, 3 (Stock number 017-001-00396-4).
49. Corn, J.K., *Protecting the Health of Workers*, ACGIH, Cincinnati, OH, 1989. pp. 35–37 and 59–65. For a discussion of development of the TLVs.
50. USEPA, *Asbestos-Containing Materials in School Buildings: A Guidance Document. Part I*, OPTS. Washington, D.C., 1979.
51. USEPA, *Guidance for Controlling Friable Asbestos-Containing Materials in Buildings*, OPTS, Washington, D.C., 1983.
52. USEPA, *Guidance for Controlling Asbestos-Containing Materials in Buildings*, OPTS, Washington, D.C., 1985.
53. USEPA, *Managing Asbestos in Place: A Building Owner's Guide to Operations and Maintenance Programs for Asbestos-Containing Materials*, PTS, Washington, D.C., 1990.
54. USEPA, *Asbestos-Containing Materials in School Buildings: A Guidance Document. Part I*, OPTS. Washington, D.C., 1979, 14.
55. USEPA, *Asbestos-Containing Materials in School Buildings: A Guidance Document. Part I*, OPTS. Washington, D.C., 1979, 13.
56a. Patton, J.L. et al., *Draft Final Report of Asbestos in Schools*, Battelle Columbus Laboratories, 1980.
56b. Logue, E.E. et al., *Draft Report on Characteristics of the Asbestos Exposure Algorithm: Empirical Distribution, Correlations and Measurement Validity*, R.T.I., 1981.
57. Findley, M.E., Assessment of the EPA asbestos algorithm, *Am. J. Publ. Health,* 73, 1197, 1983.
58. *Federal Register, Friable Asbestos-Containing Materials in Schools: Identification and Notification*, 47(103), 23370, 1982.
59. USEPA,*Guidance for Controlling Friable Asbestos-Containing Materials in Buildings*, OPTS, Washington, D.C., 1983, 3.
60. USEPA,*Guidance for Controlling Asbestos-Containing Materials in Buildings,*. OPTS, Washington, D.C., 1983, 4.
61. *Federal Register*, 52(83), 1987.
62. USEPA, *Managing Asbestos in Place: A Building Owner's Guide to Operations and Maintenance Programs for Asbestos-Containing Materials*, PTS, Washington, D.C., 1990.

On the Road to AHERA

4 Before AHERA

Confusion reigned and generated a frantic, expensive effort to remove asbestos from the schools. The Technical Assistance Program (TAP) of the Environmental Protection Agency (EPA), designed to help state and school officials voluntarily identify and correct asbestos hazards, unsuccessfully pressed schools to inspect for asbestos. EPA's "state-of-the-art" documents, intended to help school officials identify and control asbestos in buildings and provide criteria for use in deciding abatement methods for utilization in particular situations, raised more questions than they answered.

In 1982, EPA issued a school identification and notification rule, referred to as the 1982 Asbestos-In-Schools Rule. It required school officials to inspect all school buildings for friable materials to determine if asbestos was present, to keep records of findings, to notify employees if they found asbestos and where it was located, to provide maintenance and custodial employees with a guide for reducing asbestos exposure, and to notify parent-teacher associations or parents directly of inspection results. Despite having both TAP and the Asbestos-in-Schools Rule on the books, Congress enacted the Asbestos School Hazard Detection and Control Act. The Congress reasoned that no systematic program to identify hazardous conditions in schools or to remedy them existed. The Department of Education, rather than EPA administered this act. It provided for financial assistance to detect and abate asbestos and establish a program to control asbestos in schools.

THE GAO REPORT

In 1981, James Florio, Chairman of the Subcommittee on Commerce, Transportation and Tourism, Committee on Energy and Commerce, and George Miller, Chairman, Subcommittee on Labor Standards, Committee on Education and Labor, House of Representatives, requested the General Accounting Office to review the progress of federal efforts to reduce asbestos in schools. Congress requested the GAO to assess: (1) the impact of TAP in stimulating state and local action to correct asbestos problems in schools; (2) other solutions EPA took or could have taken to protect school children from asbestos; and (3) compliance by states, EPA, and the Department of Education with the Asbestos School Hazard Detection Control Act.[1] Authors of the GAO document, *Asbestos-In-Schools: A Dilemma*, interviewed representatives of the Occupational Safety and Health Administration (OSHA), the Consumer Product Safety Commission, the Department of Justice, the National Cancer Institute, the Asbestos Information Association of North America, and the Environmental

Defense Fund,* to discuss hazards in schools and how to deal with them. GAO reviewed actions taken in ten states and the District of Columbia.[2] The document GAO published declared TAP a limited success, but noted that although TAP did stimulate some inspection and abatement activity it did not institute requirements for states or school districts. TAP did not include definitive criteria for determining if the asbestos found during an inspection warranted corrective action or what type of action should be taken. The key question remained unanswered: When is asbestos hazardous enough to warrant abatement? The EPA algorithm in the 1979 guidance document listed eight factors to consider when assessing asbestos exposure in schools: (1) condition of material; (2) water damage; (3) exposed surface area; (4) accessibility; (5) activity and movement; (6) air plenum or direct air stream; (7) friability; and (8) asbestos content. However, this algorithm was found unreliable by a subsequent study.[3] None of the 11 states interviewed by GAO initiated its asbestos-in-schools activities because of TAP. "State officials repeatedly cited publicity about asbestos problems as the reason they began their program."[4] Public and media pressure in five school districts (Sacramento Unified, San Francisco Unified, Philadelphia, Pittsburgh, and Decatur) also contributed to abatement decisions, according to the GAO report. For example, the asbestos coordinator for the Philadelphia School District told GAO that asbestos-containing material (ACM) was found on surfaces such as walls and ceilings in about 18% of all buildings. None was considered hazardous; but because of public pressure, all asbestos material would be removed or encapsulated. Similarly, the Chief Construction Inspector for the Pittsburgh School District stated that based on EPA's guidance documents, district officials concluded that no asbestos hazards existed. However, they removed asbestos in three schools and encapsulated it in another, because of pressure from media, parents, and teacher groups.[5]

The Environmental Defense Fund viewed TAP as an insufficient response to the problem of asbestos-in-schools. In 1979, they sued EPA to impel it to begin rule making. EPA voluntarily reversed its original position to avoid rule making and issued an advance notice of proposed rule making on asbestos in schools in 1980.[6] The proposed rule became final on May 27, 1982. The new rule according to the GAO report, "lacked hazard criteria."

The GAO came to the following conclusions: (1) TAP was partially successful, providing some information and some inspection and abatement activity; (2) data on inspections were incomplete, and actions taken were not always as a result of TAP; (3) the program's effectiveness was limited by its voluntary nature and lack of criteria to define when asbestos is hazardous and needs abatement; and (4) lack of definitive federal criteria resulted in state and local use of different criteria for abatement decisions. The GAO further concluded that without criteria school officials will either continue to overreact and spend money needlessly, or underact and expose school occupants to hazardous asbestos conditions.[7]

---

* The EDF, a private nonprofit activist environmental organization, pressed for action to remove asbestos from schools as early as 1978. EDF believed evidence demonstrated a clear association between asbestos and cancer and that asbestos in schools would increase cancer rates among school children and posed an unreasonable risk of injury. EDF filed a petition asking EPA to initiate rule making to control asbestos emissions in schools.

The GAO conclusions regarding the Asbestos School Hazard Detection and Control Act of 1980 are just as dismal. For example, GAO said the Act had little impact on state and local activities for asbestos in schools. Its primary purpose, to provide funds for detection and correction of hazardous asbestos in schools, was not achieved. Information about when asbestos poses a hazard in schools, and when to abate was not provided.[8] In fact, none of the programs in the early 1980s resolved the dilemma school officials faced when trying to distinguish between hazardous situations needing correction, and those which presented insignificant risks. Lack of criteria led state and local officials to make decisions varying from no action in one school district to total asbestos removal in another, without assurance that school occupants were adequately protected.

EPA AND THE DILEMMA CREATED BY FEDERAL POLICY

While EPA helped to create a panic level of concern about asbestos by neglecting to specify unsafe levels of it, parents forced removal of asbestos. The following scenario was not an unusual occurrence. Similar activity happened in various locations. This case is presented to illustrate the dilemma created by federal policy for asbestos in schools.

In 1983, a mother of two children mobilized other parents to set up a committee to pressure school officials to remove asbestos in Romulus, Michigan. Parents spoke at school board meetings, sent flyers to the homes of students, operated an information booth at the city fair, and showed a film about asbestos in each elementary school in the school district. The Romulus school board buckled under the pressure. They spent $100,000 to remove asbestos from the high school ceilings, even though voters rejected bond issues to finance the project. The mother who mobilized others was quoted as saying, "If I had to do it all over again, I would picket that school until they shut it down. People were not listening to our arguments, but my son was going in there every day and breathing that air. You don't fool around with something like asbestos."[9] School officials literally did not know what to do. They were pressured to act without facts. The hard questions were: (1) Should they get rid of the asbestos?; (2) How to pay for the work of removing asbestos?; and (3) Was there truly a risk to occupants of the school building? These questions remained unanswered, but fear and panic motivated parents to insist that asbestos be removed, often incurring high costs.

RUCKELSHAUS INTERVIEW

Two journalists, Mike Wowk and Michael J. Bennett, interviewed former EPA Administrator, William Ruckelshaus, on February 15, 1985. Ruckelshaus was EPA Administrator from 1970 to 1973 and again during 1983 and 1984. Ruckelshaus' answers to some of the journalists' questions will help explain the EPA's policy decisions, as well as the reaction of parents and school administrators.

Question: What was the real reason behind the school rule?

Ruckelshaus: It was simply a question of who was going to pay for it, who was going to pay to clean it up ... The approach that was adopted by the administration before I got here was the burden was to be borne by the people living in the school districts. That's what the fight's all over. Who's going to pay for it?

Question: An EPA lawyer told me the theory behind all of this is to whip the mothers up into a vigilante mob to storm the school committee to do something?

Ruckelshaus: You could put it that way. Obviously the purpose of the approach was to bring pressure on the school boards to correct whatever situation they found.

First of all, do the inspection. Do you have a problem? If you don't, don't worry about it. If you've done the inspection and you find you have a problem that ought to be addressed, then you should notify the parents.

If you notify the parents, the parents will bring pressure upon you to correct it.

Question: The law requires schools to inspect for asbestos. Why not write something into it that would have required some action on their part?

Ruckelshaus: Well you could.

Question: Well, why wasn't that done?

Ruckelshaus: The agency could do it ... There's the authority in now for the agency to order action. The theory was if the federal government ordered the action, what would rapidly follow was a requirement that the federal government pay for whatever action was ordered. That's what scared them.

And if the federal government had to pay for it, you could just bet the cost would be several orders of magnitude greater than if you had some balance with whatever risk reduction you wanted to achieve ...

If you removed the requirement of paying from the people who are going to benefit, which is exactly what we've done with these hazardous waste sites, the demand then is for zero risk. And the cost just goes straight up because it's a cost-free benefit to them. Why shouldn't they demand zero risk?

And so the theory is you've got to keep the cost and the benefit in the same person. And then you can force them to come to grips with what sort of reductions make sense, what is reasonable to do under the circumstances ...

Now that was the theory behind trying to force the local school boards to do the inspection, notify the parents and then the parents would come in and make that kind of balance.

Question: So that was a deliberate decision.

Ruckelshaus: Absolutely.

Question: So what you're doing is essentially putting cost-benefit analysis on the agenda of every community in the country that supports a school system?

Ruckelshaus: Sure, absolutely. We try to present to them what we think the risk is, what can be done to correct it — force them to go and find out if they have any risk and then force them to make a judgement as to what you should do about it.

Now the question of whether you panic them or whether the whole thing is going to cause more harm than good, frankly. I think that's there any way you do it. I don't care whether the federal government does it or the local government does it. I think you've got the same problem.

By putting the onus on the person to make the judgement as to how much risk they want to reduce as long as the same guy is going to pay for it, it's in my view sensible. You take that away from them, which is exactly what we've done in the Love Canal kind of situation, boy, they just demand zero risk — they want the stuff out of there. "Don't tell me how much it costs. We don't care." And why should they?

Question: Dr. Bob Sawyer of the Mt. Sinai Hospital faculty (in New York) has said flatly that the asbestos policy is going to kill people.

Ruckelshaus: Well, you know, whenever we have a problem like we have with this asbestos all over the country … and we have no government apparatus to grapple with it, the potential for people panicking … is there.

Any program that we would have had — whether the government moved into it, with more money, with a training program, with a contractor — certification program — still would have been faced with some people deciding, once they had discovered that they had asbestos in their buildings or their schools, to go forward without waiting for the government program to take hold. That potential is there.

Our people were terribly worried, at EPA, about whether we would have people coming in and trying to remove asbestos before they knew what they were doing, and potentially causing a worse problem then was there before. That potential is there — there's no question about it.

What we have consistently recommended is that they be very careful about how they proceed — that they do proceed, but that they be careful about how they do it. And we provided them guidance.

And the question of whether or not they could go ahead and create a worse situation, the potential is there. There is no doubt.

Question: Malcolm Ross of the U.S. Geological Survey is saying not just that there's a panic, but there's very good reason to argue whether or not the panic is justified.

Ruckelshaus: I've listened to him have a lengthy discussion with our scientists. I sat in a room for a couple of hours and listened to them discuss the whole issue of blue and white asbestos and which one was really carcinogenic and which one wasn't.

Our people believe, our scientists believe very strongly, that while one — I can't remember which one — is far more carcinogenic than the other, they're both carcinogenic. Those kinds of uncertainties, whether it's EDB, DDT or asbestos, are all part of this equation.

There is enormous scientific uncertainty in this field ...

That isn't to say this guy (Ross) is wrong, and these other scientists are right. I mean there isn't any right or wrong here. It's just a question of where the consensus comes out. The consensus came out not supporting his position as he was espousing it within the government. He may prove to be right ...

It is one of the most frustrating parts of the job of being Administrator of EPA. It is that when you're operating in these areas of enormous scientific uncertainty, even though you try to make it as clear as you can that this uncertainty exists, people continue to take a number out of risk assessment methodology and write it in stone, as if that number were exactly how many deaths were going to occur over a particular period of time. And we don't say that.

We indicate that this is the upper bound of the risk; the lower bound can be all the way down to zero. But based on the risk as we know it and what can be done about it, here is our judgment on how we ought to proceed.

Question: Is there is rational standard of asbestos exposure?

Ruckelshaus: Our scientists have consistently told me we don't know how to set an ambient air quality standard for asbestos that makes any sense. We don't know how to do it.

So what we have suggested instead are these guidelines or approaches that can be taken where you find problems of a certain magnitude in a school, commercial building, wherever you find it. And the guidelines, in effect, take the place of the standard.

With asbestos there is always some residual risk ... where a fiber can provide a risk if somebody lives long enough.

Question: Do you think environmentalists exist not to be satisfied?

Ruckelshaus: That's true, but those people don't necessarily dominate the bureaucracy in EPA. One of the problems in this whole field ... (is) the paranoid style of American politics which so dominates the thinking.

The Congress writes these laws as though the executive branch is not to be trusted in implementing them ... They give us deadlines in which to act, which in most cases, are unreasonable. They set standards for achievement, usually zero risk, which we can't achieve either.

And then if we don't do that we get a citizen's right-to-sue provision and turn the whole thing over to the courts. I think that's a terrible way to proceed ...

What we've done is to divide the responsibility between the Congress and the executive branch and the courts, and in effect, locate the power to act nowhere. And we just bounce back and forth between these various branches of government — without, in my view, providing the clear criteria for balancing risks and benefits and costs that the agency ought to be balancing ... Instead, all of this responsibility is mixed between the Congress and the executive branch.

Getting the Congress to respond differently than we have traditionally in dealing with these problems is something that just defies me."[10]

In retrospect, the interview with Mr. Ruckelshaus appears to be an apology or rationalization for a poorly conceived, poorly-executed public health policy to deal with asbestos in schools. According to this interview, EPA responded to public pressure, largely of its own making, with instant solutions to complicated scientific, political, and economic issues. The agency unthinkingly and inexpertly juggled protective strategies, political realities, and economic issues while unsure of how or even why they were proceeding. They ignored science. The British refer to this as muddling through. The result was costly. Public sentiment informed by EPA and carried to a feverish pitch shaped the policy, and in the end because of the manner in which EPA presented and defined the issues of asbestos in schools, inappropriate public policy resulted. Ruckelshaus, whether he meant to or not, portrayed EPA as an agency unable to utilize scientific knowledge or perhaps unaware of scientific knowledge gained from years of dealing with asbestos in the workplace. OSHA, in direct contrast to EPA, published a risk assessment and a proposal for regulation of asbestos — and an airborne standard in 1986.[11]

The requirement that all schools test for the presence of ACM and, if found, to notify teachers and parents compounded public concern, because EPA did not provide guidelines for schools to utilize to determine the extent of hazard, if any, and what action to take. In fact, they led parents and school boards to conclude that ACM posed a high risk of cancer and thus to opt for removal of the material. EPA's policies, rather than helping school officials to make rational decisions regarding asbestos in schools, succeeded in creating problems that may not have previously existed. Many school systems unnecessarily removed ACM at great economic cost to the school system. They would later learn that removal of asbestos posed a greater risk than leaving it in place because of the likelihood of elevated exposure levels during and after removal.[12]

And there was the issue of the cost of removal alluded to by Mr. Ruckelshaus.* As early as 1980, Congress in the Asbestos School Hazard Detection and Control Act, directed the Attorney General to: "conduct an investigation to determine whether, by using all available means, the United States should or could recover from any person determined by the Attorney General to be liable for such costs, the amounts expended by the United States to carry out this act (federal grants and loans to detect, remove and contain friable asbestos in the schools). Within one year after the effective

* The price of removal would eventually be tens of billions of dollars. The Attorney General's report leads me to believe there was more concern about economics than public health.

date of this act, the Attorney General shall submit to the Congress a report containing the results of the study, together with any appropriate recommendations."[13]

THE ATTORNEY GENERAL'S REPORT

The resulting 231 page report of this investigation by the Attorney General's office came to a number of conclusions and based its recommendations on the conclusions. The lawyers who wrote this report did not differentiate between occupational and non-occupational exposures. They relied on court cases filed by workers and noted that workers were allowed recovery under the tort theory of strict liability, predicated on "failure to warn" and "failure to test." They noted that despite the danger, industry was "silent with respect to the dangerous relationship between asbestos and cancer," and cited the case of Hardy vs. Johns-Manville Sales Corporation. They also noted that a convincing case could be made based upon industry documents produced in litigation, stating that certain industry officials actively sought to obscure data linking asbestos and fatal diseases.[14] The Attorney General's Report neglected to discuss scientific issues, for example, the extent of health risk to school children and other occupants of schools buildings and the difference between occupational and non-occupational exposure (i.e., high and low dose). They assumed if fibers are released, danger exists. The authors of the Report note that asbestos fibers are microscopic and submicroscopic. On the same page they wrote, "The preparers of this report have personally observed in schools visited, evidence of fiber release."[15] That's hard to do if the fibers are microscopic. Whatever did they mean?

The Attorney General's report said that most of the nation's schools did not contain asbestos, although those with asbestos ranged geographically from Massachusetts to California. On the other hand, they said the square footage of coverage could be extensive and make the cost of removal or containment expensive. They went on to note that the Cinnaminson Township Board of Education in New Jersey alleged that it spent over one million dollars to deal with asbestos problems in three schools. To illustrate the extent of the problem they cited approximately 20% of New York City Public Schools and 10% of New Jersey schools as having been found to contain "asbestos materials" in student-use areas.[16] Local and state government had the responsibility of abatement of asbestos in schools because Congress had not appropriated funds under the act to make federal grants and loans. EPA did not promulgate a rule requiring school authorities to take corrective action based on their conclusions that identification of hazard will provide enough information for local school districts to take action.

Under the section of the Report, "Findings and Recommendations," the report stated:

> The parties most likely liable are the asbestos processors and manufacturers and manufacturers of the asbestos spray-on products. It will be necessary to establish that the known danger to asbestos workers should have caused these parties to: (1) test to determine whether friable asbestos could be hazardous as a result of indoor environmental exposures; and (2) warn that asbestos fibers had caused deaths and injuries in occupational settings, and if released from asbestos products, could prove harmful as

> a result of indoor environmental exposures ... However, to establish liability, it will also be necessary to prove either as a matter of law or by trial of a factual issue, that friable asbestos is hazardous."[17]

The report stressed the fiscal impact on federal tax payers, if federal funds were to be provided, unless liability was imposed on asbestos manufacturers by a statute similar to the Superfund or the Black Lung programs. The report suggested that school authorities faced with "substantial expenditures" for removing or containing friable asbestos should "as a matter of the utmost urgency" consult with a qualified counsel to determine if they should file litigation "as at least three school districts have already done." They stated that the hope of federal assistance could obstruct, not aid recovering funds.

The bottom line was that school districts with substantial abatement expenditures should be able to recover from the manufacturers and sellers of asbestos products who they believed did so without warning about the dangers of breathing asbestos. That was their advice. The conclusion is as follows.

> Litigation, but by school authorities rather than the federal government, should be quickly investigated by school authorities and their counsel as one potential means of reducing the fiscal impact on taxpayers of abating asbestos hazards in the schools. The federal government should support local school authorities in such litigation, but should not bring such actions on its own.[18]

As can well be imagined, this report opened the legal floodgates, so to speak. It reminded school authorities of the financial potential of suits against asbestos manufacturers and recommended such action. The rush to judgement and the endorsement of overly simple solutions by the Department of Justice not only equaled the opportunistic attitude of EPA, it surpassed it.

SCIENCE: THE GENERAL ACCOUNTING OFFICE AND THE ATTORNEY GENERAL'S REPORT

Both the GAO and Attorney's General's reports ignored existing scientific information, as EPA had done. They did not incorporate determination of the extent of risk in their reports, and were unduly influenced by the occupational exposures, which were orders of magnitude higher than those experienced by school children. Apparently they did not base conclusions on scientific reasoning or evidence, but on assumptions that a hazard existed. Perhaps that explains why the reports had such a negative influence and led to inappropriate, poorly performed removal of ACM from schools. The removals often worsened conditions in school, increasing the asbestos-in-air concentrations and creating a hazard for workers who removed the asbestos.

THE SERVICE EMPLOYEES INTERNATIONAL UNION

The Service Employees International Union (SEIU), also unhappy with asbestos policy as it stood, and "perceiving a high rate of lung cancer among school building service and maintenance members," declared that asbestos hazards in schools were widespread. In 1983, they commissioned a nationwide study of asbestos in schools which found 3.24 million children and 648,000 school workers at potential risk.[19] The following is a table based on a Service Employees International Union document entitled "Background Information on EPA Regarding Asbestos in Schools." It is clear from the document that SEIU had problems with EPA actions concerning asbestos in schools. They influenced Congress to address asbestos in schools by pressing vigorously for the Asbestos Hazard Emergency Response Act (AHERA). Table 1, while definitely an SEIU biased view of their role in how the union pressed to protect its members, offers another insight into why EPA policy in the early 1980s was not working.

SEIU believed that asbestos in school buildings posed a potential threat to its members when they petitioned EPA in November 1983, and later filed a lawsuit in January, 1985. They wanted EPA to: (1) establish standards for the performance of asbestos abatement activities, including standards for the protection of persons performing such activities; (2) establish standards for determining when friable ACM in schools are hazardous; (3) establish requirements for corrective action in schools when friable ACM are determined to be hazardous; and (4) establish requirements for inspection and abatement of friable asbestos-containing materials in public and commercial buildings.[21] SEIU also believed EPA had taken the "path of least resistance" by not adopting an asbestos standard similar to that of OSHA.[22]

THE WALLS AND HALLS OF TROUBLE — THE AMERICAN ASSOCIATION OF SCHOOL ADMINISTRATORS

In 1978, an article published in *The American School Board Journal* entitled, "Asbestos in Schools: Walls and Halls of Trouble"[23] alerted school board officials to the coming issue of asbestos in schools. This early call to attention stated that asbestos could be solved by "astute school boards." The author suggested that school boards ask certain questions, and counseled a cautious and questioning approach to the problem. "If exposure to asbestos can cause cancer and if the walls and ceilings of your school are covered with asbestos, does that mean that your students and employees face a health danger just from being inside your school buildings? That question can't be answered with a simple "yes" or "no"; it requires replies containing "maybe" and "possibly" and "it all depends." Your responsibility in this matter, however, is clear-cut: to begin asking key questions. Do your schools contain asbestos? To what extent is that asbestos a health hazard to students and employees? How can the health danger be eliminated?. What's it all going to cost?"[24] In 1979, when EPA issued proposed rules for elimination of asbestos from school buildings, the American Association of School Administrators (AASA) called for an independent scientific panel to investigate the hazards of asbestos in schools. In a letter to Patricia Harris, Secretary of Health Education and Welfare, the AASA stated their concerns about health hazards but noted that in the case of asbestos, levels that might cause health hazards are unknown. The letter said, "To date there has been no study to indicate that asbestos of the kind in school buildings has presented a hazard to

TABLE 1
SEIU Activity Concerning Asbestos in Schools[20]

Date	Service Employees International Union Activity
March, 1983	SEIU locals perceive high rate of cancer among school building service maintenance members. Members claim asbestos hazards in schools widespread.
March–June, 1983	SEIU commissions first nationwide study of asbestos in schools that finds 3.24 million children and 648,000 school workers potentially at risk.
September, 1983	SEIU studies in detail why asbestos hazards in schools are not being abated. They find 2 main reasons: (1) lack of EPA rule requiring safe and complete abatement of hazardous asbestos; and (2) lack of adequate funding to schools district.
October, 1983	SEIU worker education pamphlet, "What Every School Worker Should Know About Asbestos."
November, 1983	SEIU (1) launches campaign to seek federal funds from Congress, and (2) petitions EPA (under TSCA) for rules requiring abatement of hazardous asbestos in a safe and complete manner.
January, 1984	SEIU meets with Ruckelshaus — he is noncommittal.
February, 1984	EPA CLAIMS SEIU petition granted — say will consider issuing regulations for safe and complete abatement. Public comments requested and one public hearing scheduled for May 7, 1984.
February, 1984	SEIU commissions nationwide study of state legislation on asbestos in schools. Report entitled, "Filling the Gaps" concluded approximately 20 states enacted various legislation, but not one has done an adequate job. Piecemeal state-by-state approach will assure asbestos will remain in many schools for decades.
April, 1984	SEIU forms broad-based Asbestos-in-Buildings Technical Advisory Committee (composed of organizations and individuals representing children, parents, workers, asbestos abatement contractors, universities, asbestos abatement health professionals, and boards of education). Committee drafts nine provisions for minimal requirements for EPA asbestos-in-buildings rule.
May–June, 1984	Four hearings held. Washington, D.C., San Francisco, Boston, Chicago
July, 1984	EPA releases preliminary Survey "prompted by SEIU petition." Find 30,800 schools with friable asbestos potentially exposing 15 million children and 1.4 million workers.
August 13, 1984	SEIU sends letter to EPA requesting a time table for issuance of abatement rules.
August 15, 1984	EPA responds to SEIU letter without a time table.
September 6, 1984	SEIU sends letter to EPA requesting time table.
September 11, 1984	SEIU files suit in U.S. District Court to force EPA to release a time table for when SEIU November, 1983 demand will be addressed.
January 29, 1985	SEIU files suit in U.S. District Court seeking review of EPA's decision not to grant all points of SEIU November 1983 petition. SEIU believes EPA decision at odds with public hearing record, EPA internal memo, and EPA school survey.

TABLE 1
SEIU Activity Concerning Asbestos in Schools[20] *(Continued)*

Date	Service Employees International Union Activity
February–May, 1985	SEIU reviews EPA internal documents which SEIU believes support need for rules. Evidence in deposition of EPA officials says SEIU shows EPA's past efforts may be killing more people than it is saving because of haphazard abatement actions.
June, 1985	The Environmental Defense Fund and a number of other organizations sign on as co-plaintiffs to SEIU lawsuit.
June, 1985	At a Congressional hearing, John Moore, EPA Assistant Administrator for Toxic Substances and Pesticides claims that the SEIU petition requests have merit, but that EPA denied the request because of lack of agency resources.
August, 1985	National Governors' Conference adopts a resolution supporting the SEIU petition requesting EPA regulations.
January, 1986	SEIU begins drafting Federal legislation with staffers from U.S. Representative James Florio's office to cover asbestos in schools.
Spring–Summer, 1986	Congressman Florio introduces legislation jointly sponsored by Senator Stafford.

children. Most studies have dealt with workers or miners handling asbestos who were subjected to much higher exposures."[25] AASA wanted to determine whether a hazard really existed; and if it did exist, how serious was it? They also desired a method to assess the asbestos problem within school buildings. After a meeting held under the auspices of the National Institute for Environmental Health Sciences (NIEHS), the AASA believed the jury was still out over whether low-level exposure to asbestos poses a health hazard.[26] They also voiced strong opposition to EPA's announced proposed rule making in the Federal Register of September 20, 1979 requiring schools to take immediate action. AASA expressed concern that the rule would place undue burden on schools and unnecessarily panic school people, students, parents, and the community.[27] The *American School Board Journal,* after discussion of EPA's proposed regulations asked, Who is going to pay for all this? They also answered the question: "The answer should come as no surprise: You, mostly although the federal government probably will kick in a nominal amount. New legislation earmarks $22.5 million for an asbestos detection program over the next two years: The law also sets up a $150 million interest-free loan program over the same period for asbestos hazard control. In August, however, Congress had not yet appropriated funds for the program."[28]

But reason did not prevail. Just a few years later in 1984, the *American School Board Journal* was publishing articles with the following titles: "Sidestep Asbestos Hysteria: Explain the Risk (and Your Response) to the Public,"[29] and "Asbestos — The Clock Ticking in Your Schools, and Inaction Could Prove Devastating."[30] In the same issue of the *Journal,* calm resolution of the problem was suggested, for example, educating the parents, calming them, not responding with hysteria to hysteria. Moreover, the article told of the need for community support and funding and noted that by the end of the summer of 1983 in Montgomery County, Alabama where the "reasoned" approach was taken, more than one million dollars was received from the state, as well as an additional $2.4 million in bonds. By the end of 1983, 360,000 square feet of ACM had been removed from 17 schools in Montgomery County, Alabama.

Advice in the second article was to act to remove. "If you think the EPA is making idle threats, think again. Under the direction of old/new Director, William Ruckleshaus, EPA has become more vigilant and is going after asbestos in schools as one of its ten top priority items. EPA's design in these aggressive actions is to force schools into compliance through pressure from the community, according to Connie DeRocco, a specialist in EPA's asbestos enforcement program. It stands to reason that school board members would tend to choose compliance over the chance that a concerned parent might blow the whistle on the school's negligence."[31] Clearly, by 1984, although the school administrators had earlier counseled moderation and intelligent questioning of the issues, the issues had become highly emotional and reason no longer prevailed.

ASBESTOS AND LEGAL ISSUES

Furthermore, the legal issues began to dominate all others. In accordance with GAO advice, attorneys advised schools to consult their school attorney and devise a legal action plan. Plaintiff's lawyers sought to educate school boards and administrators on legal questions. The most prevalent question was: Can you recover the cost of removing asbestos from manufacturers of asbestos products? For example, "Although Congress enacted the Asbestos School Act, it has thus far failed to appropriate funds to help meet the expense involved in identifying and abating the asbestos hazard. It is, therefore, essential that local school boards fully and immediately address the potential asbestos hazard and when costs are necessarily incurred, vigorously pursue their legal rights to force companies that knowingly placed asbestos in the schools, to pay for the substantial cost of abating the hazard."[32]

School boards and administrators received mixed messages. At first cautioned to carefully assess the problem and approach solutions rationally, they soon succumbed to pressure to remove asbestos and seek redress, i.e., monetary compensation in the courts. They moved from concern for public health to concern for economics. It was rather difficult not to respond to the legal temptation, given pressure from the federal government, i.e., EPA, the Attorney General's Office, plaintiff lawyers, the SEIU, EDF, and irate parents. And the most relevant question, "Was asbestos in schools damaging to health of school children," apparently got lost. The outcome was that schools did not make informed decisions about asbestos management. Communications were unclear and inconsistent. Diverse organizations and businesses promoted their own information and concepts of the issue. For example, companies who earned billions of dollars removing asbestos from school buildings. School officials made significant public health and financial decisions, all within a context of conflict and misunderstanding.

PUBLIC AWARENESS OF ASBESTOS IN SCHOOLS

In the period 1970–1985, from the early NESHAP rules of the Clean Air Act through the beginning regulation for asbestos under the Toxic Substances Control Act, EPA attempted to raise public awareness about the hazards of asbestos. They directed efforts toward parents, state health and environmental agencies, and especially local education agencies (LEAs). EPA's position was that the removal or remediation of asbestos in school buildings should not become a federal bail out program but that those who benefitted should pay, i.e., local school systems. EPA intended, with the 1982 School Inspection and Notification regulation, to increase protection of school children by requiring the identification of friable asbestos which they thought would lead to voluntary safe work practices. But it did not. The regulation which sought notification of parents increased pressure from parents on local education agencies to respond to asbestos by removing it. EPA itself estimated low compliance with most aspects of the regulation and most inspections performed were done poorly by people with little or no training.[33] "Whatever the shortcomings of the Inspection and Notification Rule, it had a significant communications impact. The perceived threat to school children appears to have increased public awareness of asbestos hazards. The EPA's 1982 rule brought the asbestos problem home to millions of parents and school officials."[34] The 1984 Congressional passage of the Asbestos School Hazard Abatement Act kept the issue alive, directing the agency to provide financial assistance to needy schools.

In this same period of time, 1970–1985, EPA Guidance documents cited earlier, *Asbestos-Containing Materials in School Buildings (The Orange Book)* and *Guidance for Controlling Friable Asbestos-Containing Materials (The Blue Book),* focused on health hazards and health effects and made people aware of potential harm. The guidance documents also led school officials to conclude that removal of ACM from school buildings was the best and "safest" course of action and to neglect other options, for example, management of in place asbestos or encapsulation of asbestos. EPA emphasized removal as the primary means to control asbestos risk.

Perhaps the most influential message came from the United States Congress. Apparently the perception in Congress was that asbestos in schools was an environmental emergency and warranted a political mandate from them to do something about this "national emergency." Congress could control EPA in two ways, by passing laws that defined the agency's authority and responsibilities and by allocating funds the agency needed to function, which they did.

REFERENCES

1. General Accounting Office, *Asbestos in Schools: A Dilemma*, U.S. General Accounting Office, Washington, D.C., Aug. 31, 1982, 4.
2. General Accounting Office, *Asbestos in Schools: A Dilemma*, U.S. General Accounting Office, Washington, D.C., Aug. 31, 1982, 5.
3. Findley, M.E., Assessment of the EPA Asbestos Algorithm, *Am. J. Publ. Health,* 73, 1179, 1983.
4. General Accounting Office, *Asbestos in Schools: A Dilemma*, U.S. General Accounting Office, Washington, D.C., Aug. 31, 1982, 10

5. General Accounting Office, *Asbestos in Schools: A Dilemma*, U.S. General Accounting Office, Washington, D.C., Aug. 31, 1982, 15.
6. *Federal Register* 45/61966. Sept. 17, 1980.
7. General Accounting Office, *Asbestos in Schools: A Dilemma*, U.S. General Accounting Office, Washington, D.C., Aug. 31, 1982, 18.
8. General Accounting Office, *Asbestos in Schools: A Dilemma*, U.S. General Accounting Office, Washington, D.C., Aug. 31, 1982, 27.
9. *Detroit News*, Sunday, March 3, 1985 (by line Mike Wowk).
10. *Detroit News*, William Ruckelshaus interviewed by Mike Wowk and Michael J. Bennett. March 5, 1985.
11. *Federal Register*, Vol. 51, No. 119, Rules and Regulations, June 20, 1986.
12. Burdett, J.D., Jaffrey, S.A.M.T., and Rood, A.P., Airborne Asbestos Levels in Buildings: A Summary of U.K. Measurements, in *Non-Occupational Exposure to Mineral Fibres*, Bignon, J., Peto, J., and Sarracci, R., Eds., IARC No. 90, International Agency for Research on Cancer. Geneva, 1989, 277.
13. U.S. Department of Justice, Land and Natural Resources Division, *The Attorney General's Asbestos Liability Report*, Washington, D.C., Sept. 21, 1981, i.
14. U.S. Department of Justice, Land and Natural Resources Division, *The Attorney General's Asbestos Liability Report*, Washington, D.C., Sept. 21, 1981, ii.
15. U.S. Department of Justice, Land and Natural Resources Division, *The Attorney General's Asbestos Liability Report*, Washington, D.C., Sept. 21, 1981, iii.
16. U.S. Department of Justice, Land and Natural Resources Division, *The Attorney General's Asbestos Liability Report*, Washington, D.C., Sept. 21, 1981, iii.
17. U.S. Department of Justice, Land and Natural Resources Division, *The Attorney General's Asbestos Liability Report*, Washington, D.C., Sept. 21, 1981, iv.
18. U.S. Department of Justice, Land and Natural Resources Division, *The Attorney General's Asbestos Liability Report*, Washington, D.C., Sept. 21, 1981, vii.
19. Service Employees International Union, *Background Information on EPA Regarding Asbestos in Schools*, Undated document prepared by SEIU.
20. Service Employees International Union, *Background Information on EPA Regarding Asbestos in Schools*, Undated document prepared by SEIU, 1.
21. *Federal Register*, Vol. 50, #134. July 12, 1983, 3. (40 CFR Part 763).
22. Letter from John J. Sweeney, (President of SEIU) to Document Control Officer (TS-739) OTS.EPA, Room 209, Washington, D.C., Sept. 10, 1985.
23. Levin, D., Asbestos in Schools: Walls and Halls of Trouble, *Am. School Board J.*, Nov. 1978.
24. Levin, D., Asbestos in Schools: Walls and Halls of Trouble, *Am. School Board J.*, Nov. 1978, 29.
25. D.C. Dateline, *The School Administrator*, 36(8), 1979.
26. D.C. Dateline, *The School Administrator*, 37(1), 1980.
27. D.C. Dateline, 37(1), 9, 1980.
28. *Am. School Board J.*, Sept. 1980. p. 12.
29. *Am. School Board J.*, April, 1984. pp. 36–37.
30. *Am. School Board J.*, Op. Cit. p. 33.
31. *Am. School Board J.*, Op. Cit. p. 35.
32. Speights, D.A., Kimmel, A., Stanford, E.H., and Brown, D., *Asbestos in Schools and Other Buildings: A Layman's Guide to the Legal Issues*, Source Finders Technical Reprint, Source Finders Information Corp., NJ, 1985.
33. USEPA, *Communicating About Risk: EPA and Asbestos in Schools*, Final Report of the Internal Task Force. Internal Document, Jan. 1992, 12.
34. USEPA, *Communicating About Risk: EPA and Asbestos in Schools*, Final Report of the Internal Task Force. Internal Document, Jan. 1992, 12.

5 Congress Passes AHERA

POLITICS AS USUAL

By 1985, the uncertainties associated with asbestos in school buildings led to even greater confusion than had previously existed about how to proceed with asbestos abatement or even whether or not to proceed at all. Some members of the scientific community had already begun to question whether there was risk of disease from low doses of inhaled asbestos in school buildings.[1] Nevertheless, the Environmental Protection Agency (EPA) continued to exaggerate the risk to school children's health from asbestos-containing materials (ACM) in schools by presenting worst case scenarios, and overlooking the fact that uncertainties did exist. It was politics as usual. Scientific issues and questions seldom surfaced in the popular press. High dose, low dose, scientific uncertainties, public health policy, legal liability, state and federal responsibility, and the role of EPA apparently were not part of the public discussion. Instead, pressed by political considerations, the United States Congress held hearings on EPA efforts to control asbestos hazards in schools. The agenda of Congress, one never hidden, was to write a new law, one that would include specific regulations in order to reduce exposure of school children to asbestos in schools. Congress did not question whether the new regulations would be in the interest of public health, and continued to base its arguments for new regulations on the worse case scenario: hypothetical risk estimates.

THE CANADIAN ROYAL COMMISSION ON ASBESTOS

In 1984, a Royal Commission in Canada conducted an investigation into matters of health and safety associated with the use of asbestos in Ontario, Canada. In the report entitled, *The Report of the Royal Commission on Matters of Health and Safety Arising from the Use of Asbestos in Ontario* (a four volume report) the Royal Commission came to the following conclusions:

> In dramatic contrast,* the exposure of building occupants to asbestos fibers during normal building use will be shown to be insignificant, whether as compared to the exposure of insulation workers in the past or as compared to the much lower exposures permitted by the recently adopted Ontario workplace control limits of 1.0, 0.5 and 0.2 f/cc for chrysotile, amosite and crocidolite asbestos. Studies of asbestos concentrations in building air have shown that many buildings containing insulation do not exhibit fibre levels exceeding those in the outdoor air or in buildings not insulated with asbestos.

---

* The contrast refers to occupational exposures.

Even when a building exhibits elevated asbestos fibre levels, these are still very low compared to current workplace control limits and are orders of magnitude below the level to which workers were exposed in the past. A typical building containing asbestos insulation will expose occupants to less than 0.001 f/cc of asbestos, or 1/1,000 of the current chrysotile control limit. Only a small fraction of occupant exposures in all buildings containing asbestos insulation would be as great as 0.01 f/cc of asbestos.

We will conclude that it is rarely necessary to take corrective action in buildings containing asbestos insulation in order to protect the general occupants of those buildings. On the other hand, construction, demolition, renovation, maintenance and custodial workers in asbestos containing buildings may be exposed to significant asbestos fibre levels, and may during their work, cause elevated fibre levels for nearby occupants. We will devote Chapter 10 to the problems of protecting occupants from the possible fibre release as a result of building work. This, and not the protection of building occupants in the absence of such work, is the real challenge that asbestos in buildings presents."[2]

SUBCOMMITTEE ON COMMERCE, TRANSPORTATION, AND TOURISM

The congressmen and women on the Subcommittee on Commerce Transportation and Tourism which held hearings in 1985 either overlooked the report of the Canadian Royal Commission or quite possibly remained ignorant of its existence. Or perhaps they did not believe it. The Subcommittee of the House of Representatives held two hearings on asbestos exposure. The first, entitled "EPA Efforts to Control Asbestos Hazards," met June 27, 1985. The second, Asbestos Hazard Emergency Response Act of 1986 (AHERA), met March 4, 1986.[3] Both hearings, along with their documentation, provide insight into the problems and issues EPA faced, the pressures brought to bear on the Agency, and the unanticipated results of its ongoing asbestos in schools policy and activities. The hearings also bring into focus congressional motivation and perception of risks associated with asbestos in school buildings.

The Subcommittee chaired by James J. Florio of New Jersey included 17 members. Table 1 lists committee members and their states.[4] Those testifying at the hearings of "EPA Efforts to Control Asbestos Hazards" included:

1. Alvin L. Alm, former EPA Deputy Administrator
2. Bill Borwegan, Health and Safety Director, Service Employees International Union
3. Edithe Fulton, President, New Jersey Education Association
4. Dan Guttman, Counsel, Service Employees International Union
5. Ruby King-Williams, Assistant Executive Director for Public Affairs, National Education Association
6. John A. Moore, Assistant Administrator Pesticides and Toxic Substances, Environmental Protection Agency
7. John R. Russell, Jr., P.E., Assistant Superintendent for Maintenance, Houston Independent School District

8. John J. Sweeney, International President, Service Employees International Union
9. Susan Vogt, Director, Asbestos Action Program, Pesticides and Toxic Substances, Environmental Protection Agency
10. Millie Waterman, National Vice President for Legislative Activity, National Parent Teacher's Association
11. John F. Welch, President, Safe Buildings Alliance[5]

TABLE 1
Subcommittee on Commerce, Transportation, and Tourism

James J. Florio, New Jersey, Chairman

Barbara A. Mikulski, Maryland	John D. Dingell, Michigan
Dennis E. Eckart, Ohio	Norman F. Lent, New York
Bill Richardson, New Mexico	Thomas Tanke, Iowa
Ralph M. Hall, Texas	Don Ritter, Pennsylvania
Philip R. Sharp, Indiana	Dan Coats, Indiana
W.J. "Billy" Tanzin, Louisiana	Jack Fields, Texas
Wayne Dowdy, Mississippi	Dan Schaefer, Colorado
Jim Slattery, Kansas	James T. Broyhill, North Carolina (Ex Officio)
Gerry Sikorski, Minnesota	

In his opening statement, Representative Florio said that the subcommittee met to investigate the federal government's efforts to control asbestos hazards in schools. It appears that he believed, and certainly led his audience to believe, that neither controversy nor uncertainties existed about the nature or extent of the health hazards associated with asbestos in schools when he said the following.

> We are now beginning to appreciate the threat it poses to occupants of schools and other public buildings. Federal government studies estimate that some 15 million children and 1.4 million school employees are potentially at risk. I and other members of the subcommittee continue to be alarmed by EPA's failure to act in accordance with the Congressional mandate to minimize and prevent harmful asbestos exposures. Despite ongoing criticism from parents, teachers, school workers, school administrators, and even the asbestos industry, all that EPA has chosen to do is to require that local officials inspect school buildings and notify parents and school employees of the results.
>
> The rule does not establish standards for local officials to determine when asbestos is hazardous and what measures to take once a hazard is identified, however it is identified. In fact, the rule does not mandate that school officials do anything at all after they have inspected and notified and presumably found a hazard however defined.[6]

The hearings would eventually degenerate into an EPA bashing session.

Florio did not doubt for a moment that asbestos in schools was a killer. His opening statement set the committee's agenda: to press EPA to set standards for evaluation and abatement of asbestos in Americans' school buildings. The Committee wished to impose a requirement by proceeding with rulemaking for abatement.

Florio implied that EPA's failure to promulgate rules, thus far, resulted from the Office of Management and Budget's interference and the Reagan administration's efforts to oppose funding in the 1986 budget for continued asbestos abatement in schools.[7] These hearings could be interpreted as backlash against the Reagan administration's extremely reactionary actions and attitude toward environmental protection.

The President had opposed funding in the 1986 budget for a continued program to abate asbestos in schools, even though Congress proposed an appropriation of 50 million dollars for abatement. Thus, part of the agenda for the hearings was to get money appropriated for asbestos abatement in spite of the fact that the Reagan administration had turned its back on environmental issues. Congress wanted asbestos funding written into federal law.

At the hearings, discussion on the social issue of asbestos in school buildings sounded like this. Mr. Eckart to Mr. Florio: "Mr. Chairman, I want to thank you for going forward and keeping the heat on this issue. There are four dimensions of peculiar concern to me." (I think he meant particular). "First and foremost is, what happens to our children attending school in institutions which were built many, many years ago, and so continue to expose themselves just by virtue of fulfilling mandatory attendance laws in the state to try to improve their lot in life." Eckert said he was concerned about the children in schools whose health he believed was imperiled because the, "... foot dragging that has taken place jeopardizes our children, who then increase their risk, because of the age at which they are exposed."[8] He also said he was especially concerned about the workers who have to deal with asbestos.

Congress wanted to goad EPA into action it considered both appropriate and politically correct. The scientific merit of the argument for legal requirements to remove asbestos, and the problems involved in removal or similar action, were not discussed or even mentioned at the hearings. A bystander attending the hearings could easily conclude that asbestos lurked in the halls and walls of all American schools, ready to pounce and endanger the lives of school children and maintenance workers by causing the dreaded disease of cancer. The subcommittee adjusted science to fit its political end. Congress intended to force the Reagan Administration to appropriate federal funds to remove asbestos from American schools.

The first witnesses represented EPA: John Moore, Assistant Administrator of Pesticides and Toxic Substances, accompanied by Susan Vogt, Director of Asbestos Action Program and Alvin L. Alm, former EPA Deputy Administrator. Moore presented EPA's program in the following manner. He characterized EPA's goals to minimize inhalation of asbestos in place in buildings by presenting the agency's five elements to achieve the goal. He said EPA approach included: (1) determination of presence and location of the ACM; (2) evaluation of its condition; (3) developing a plan to address the situation; (4) selecting qualified people to perform asbestos abatement work; and (5) disposing of the material appropriately.[9] Moore also said EPA emphasized strong technical assistance, rulemaking where practicable, and active enforcement of rulings.

He listed what he considered significant accomplishments of the agency. The accomplishments included:

1. the implementation of the Asbestos School Hazard Abatement Act
2. revision of asbestos guidance document
3. grant awards for state contractor certification
4. technical assistance programs
5. worker protection rule
6. establishment of three asbestos information and training centers
7. publication of an asbestos waste management guidance document.[10]

He concluded saying that the agency had made substantial progress toward implementing and expanding asbestos activities. Moore could not speak otherwise. He was, after all, Assistant Administrator for Pesticides and Toxic Substances. His job was to show how much progress EPA made toward implementing federal asbestos policy.

Congressman Norman Lent's (New York) opening statement supported the popular perception about asbestos risks, that "asbestos materials," (his use of the words) in schools placed children at "considerable risk." He said in his testimony that legislators desired, "... some uniform standard, that will magically solve the problem."[11]

Florio and Lent chastised EPA for failing to set national standards. They argued that local decision makers lacked technical expertise, often receiving bad and expensive advice from contractors or consultants. Florio asked Moore,

> Isn't it the case, you have regional asbestos coordinator surveys, the survey taken in the spring of 1984 if I recall, reveals, the survey indicates that 75% of the abatement activities that have been taken were done improperly, improperly being not in accordance with your own guidelines. You are saying, well, we can't get into regulations, we can't get into standards, we put out these guidelines.
>
> The abatement activities are not taking place as we would like them to, even the inspection, but the abatements are taking place 75% of them are not in accordance with your generalized guidelines.[12]

Lent said that compliance with existing regulations, i.e., the Asbestos in Schools Hazard Abatement Act (School Identification and Notification Rule) was fairly low, especially concerning notification requirements.[13]

A great deal of discussion occurred on the subjects of worker protection and improper and/or careless abatement activity. Both Congressmen Florio and Lent chastised John Moore for not being more responsive to the Service Employees International Union (SEIU) petition for worker safety regulations. They continuously noted that abatement activities fell short of the desired results. For example, Mr. Florio said to Mr. Moore, "you tell me that you don't know that those 75% that were inadequately done (inspections) have been corrected. That is over a year now. Are you prepared to represent to us today that the deficient abatement actions that were taken have been corrected?" Moore responded, "No I am not willing to state that. I don't think there is anyway I could do it. The 75% figure that you are quoting emanated from a discussion that was held at a national meeting with regional asbestos people involved." Florio then said, "Your regional coordinators?" Moore answered,

"Yes, EPA people involved with the question was, How well do you think abatement is going? That figure (75%) has no validity other than to say that given that the number numerically comes out to 75% a very large number of abatement activities, based on the opinion of those back in March 1984, were not being done in a completely acceptable fashion. Based on that, we are trying to do a variety of activities, many of which I have articulated here this morning, to try to reduce that number."[14]

The SEIU and EPA

John J. Sweeney, International President of the Service Employees International Union (AFL-CIO), Bill Borwegan, Health and Safety Director, and Dan Guttman, Counsel, testified at the hearing on the need for rulemaking. Sweeney said the union was already seeking a court order, before a United States District Court, to make EPA initiate rulemaking that would require corrective action when asbestos is present in schools. The SEIU maintained that EPA said it would initiate rulemaking and subsequently reneged on its promise to do so. Sweeney said that at hearings held in response to the SEIU petition, EPA did not seek rulemaking, although it knew of widespread, improper abatement, and that the program caused more harm than it prevented.

The SEIU rhetoric sounded like this.

> In this case, the false claim of a government agency is more than self serving when EPA tells the public that ongoing activities are successful, when it knows that they are not. It both discourages reinspection that might uncover hazards the EPA boasts are covering up, and encourages further abatement activity that may well be improper and increase hazard ... EPA's own contractor survey shows, what the contractor reports of EPA field officials confirm, that EPA's claim that its program is working is not only false but most pointedly ignores widespread under reporting of the presence of friable asbestos-containing materials in boiler rooms.[15]

The SEIU wanted rulemaking. No doubts existed in their minds about the hazards associated with asbestos in school buildings.

Earlier, on November 13, 1983, the SEIU petitioned EPA to initiate rulemaking concerning inspection for, and abatement of, asbestos hazards. EPA responded to the petition by holding four public meetings. Nevertheless, on September 11, 1984, SEIU filed suit against EPA for failing to complete rulemaking within a reasonable time. EPA's position in 1984 on a rule to force removal of asbestos was that a specific requirement would result in federal enforcement, which would place the schools against the federal government.

"We think it is more productive to utilize hundreds of thousands of concerned parents instead of government inspectors to stimulate action."[16] Apparently the agency wanted to direct its initiatives toward prodding school systems to act. This theme is a recurring one in agency asbestos policy.

But this was "bash EPA day" at the hearings. Everybody concerned appeared to be discontented with EPA's policy decisions and its handling of the problem of asbestos in schools. No doubt asbestos in schools was considered a problem by members of the congressional committee and the witnesses. Nobody questioned whether a problem existed. All who spoke at the hearings disagreed with, or were unhappy with, EPA's policy decisions and its handling of asbestos in schools. Those who spoke at the hearings wanted a new law which included rulemaking for how to proceed with asbestos removal. After reading the hearing transcriptions one is reminded of Rashamon. Each person, representing different constituents, interpreted the problem and solutions differently. They all agreed on EPA's inability to solve the problem.

NEA

The leadership of the National Education Association (NEA) also appeared before the subcommittee to present its point of view. Ruby King-Williams, Assistant Executive Director for Public Affairs, spoke for the group. First, she reiterated the conventional wisdom about asbestos risk. She called it a serious threat, said medical research proved that once inhaled, asbestos fibers remain in lungs indefinitely, that 15 million students and 1.4 million school employees are being exposed to asbestos, and that the health hazard to children is particularly serious since children exposed to asbestos are more likely to develop cancer than similarly exposed adults. Williams finally made her point. EPA refused to issue regulations establishing a clear standard to determine when asbestos is hazardous, and requires abatement. The NEA believed it essential for Congress to mandate that EPA promulgate regulations to establish uniform federal standards by which all schools would assess their potential asbestos hazards and take measures to permanently and effectively abate them.[17]

Millie Waterman, National Vice President for Legislative Activity of the National Parent-Teacher Association, represented the organization and urged the committee to recognize the seriousness of the asbestos problem and the need for corrective action. She urged the committee to approve legislation to mandate EPA to: (1) set procedural standards to determine if an asbestos hazard exists; (2) require abatement; (3) require certification of contractors and consultants; and (4) require periodic re-inspection and notification.[18]

The last two witnesses, John P. Russell and John F. Welch, represented the Houston Independent School District (HISD) and Safe Buildings Alliance, respectively. John P. Russell, Assistant Superintendent for Maintenance at HISD, presented his school district's experiences with its asbestos school hazard abatement program. The maintenance department had surveyed all district facilities and determined that 118 of 234 schools had friable asbestos, at least in parts of each facility. He described how HISD sought to abate the asbestos. According to Mr. Russell, the approach was a coordinated and very expensive one.[19] It should also be noted that the school district had already sued asbestos product manufacturers to recover costs after removal. Mr. Russell said he considered his school district's approach to asbestos abatement rational. First he praised EPA for the help it offered under the 1984 ASHRA law. He said,

> ... they have provided assistance in improving our minimization procedures. They have worked with our Texas Department of Public Health, the American Wall Sealing

> Institute, the Georgia Institute of Technology. All these types of actions, especially at the region six level, have been very, very helpful to us.
>
> However, without a national standard for asbestos exposure-allowable exposure to occupants, it has caused us some heartburn. Without the benefit of these standards, we have had to accelerate our program, because our school board doesn't know really how hazardous a given school is. Therefore, all schools that have asbestos have to be considered hazardous.[20]

John Welch, representing the Safe Buildings Alliance (SBA), referred to the findings of both the Ontario Royal Commission, that ACM in most buildings rarely pose a significant risk as well as the Doll and Peto report, which concluded that environmental health risk from asbestos in buildings is of an extremely low order of magnitude. Welch, in fact, was the only person during the day-long hearing who cited these two documents. He said the position of the SBA was not to adopt a "do nothing attitude, nor to overreact to the presence of asbestos and hastily undertake asbestos abatement activities and thus create problems where none previously existed. Building owners are conscious of the need to take appropriate steps to insure proper management of asbestos-containing materials." He said the SBA strongly supported worker protection rules, along with operations and maintenance programs and the need to utilize air-monitoring techniques to assesses building conditions. He said the SBA did not seek more regulation. In a conciliatory vein, he concluded, "SBA finds that widespread attention to the asbestos in buildings issue is slowly leading toward appropriate, rational responses, and that building owners and policy makers at the State are taking appropriate steps to insure that risks do not occur."[21]

Nevertheless, Congress would take inappropriate, irrational measures in the form of the Asbestos Hazard Emergency Response Act of 1987, egged on by the very hearings under discussion. And after a full day of hearings, and unabashed political posturing one thing surely became clear. No one endorsed the environmental policy of the Reagan administration or the actions of EPA. All, except the SBA, wanted a federal law to enforce asbestos policy for schools, even if nobody knew exactly its purpose and its intent.

Committee Hearings, March 4, 1986

Eight months later, on March 4, 1986, the same committee held hearings on the Asbestos Hazard Emergency Response Act of 1986. Once again witnesses and legislators chastised EPA approach to asbestos in schools which they concluded did not work. All witnesses except those representing EPA testified that EPA policy did not work. Susan F. Vogt, Office of Pesticides and Toxic Substances at EPA and David Mayer, Chief of the Technical Assistance Program at EPA disagreed with the others. Susan Vogt updated the House committee about EPA's asbestos program: "In summary I believe the agency has made substantial progress in continuing to implement our program in the last 6 months. We plan to continue to work toward our mutual goal of minimizing the inhalation of asbestos in school buildings."[22] All the same, no real discussion of the risks or of assessment of asbestos in buildings occurred;

as if scientific questions did not exist. The hearings continued as usual. For example, Barbara Mikulski (Maryland) said,

> Mr. Chairman, I am pleased to be an original cosponsor of the Asbestos Hazard Emergency Response Act of 1986. I've been urging to get asbestos out of the classrooms and putting teachers back since 1979.
>
> That's when asbestos was first identified as a serious and significant health risk to over 16 million kids and school employees.
>
> In 1986 we are learning there are over 31,000 school buildings in this country that still contain the kind of asbestos that is most easily inhaled and therefore, in some ways, the most dangerous. EPA has failed to respond to this problem. It is a frightening situation, and they have left us in a situation where we have many questions.
>
> EPA has not solved the problem. EPA has contributed to the problem. They have instructed school officials to tell parents that asbestos is in school, but they have not been able to say how safe is safe. They have not told school personnel how to solve the problem by removing asbestos in a way that does not turn America's schools into a new Superfund site.
>
> I'm glad that your legislation requires that we develop proper inspection for asbestos, identifying what constitutes hazards and how we can get rid of asbestos without increasing the danger of exposure.
>
> By EPA's own estimates as much as 75 percent of all asbestos cleanup work being done in our schools is being done incorrectly. We must act now and we must act quickly.
>
> The purpose of this legislation before us today is to make sure the Federal government does what must be done to protect our school children, protect our teachers and other school employees. People have a right to be heard and we are doing that today.
>
> People have a right to know about the hazards. But they also have a right to know what to do about the hazards.[23]

It is not clear that either EPA or its severe critic, Representative Mikulski, understood the problem. Mikulski and the other committee members wanted a solution without an assessment of the problem. Once again, the scientific knowledge that could have been utilized to assess the problem was ignored and politics as usual prevailed.

Conventional Wisdom

The following quote sums up the conventional belief in Congress and elsewhere about asbestos in schools, and the role of EPA.

> Mr. Speaker, I appreciate this opportunity to lend my support to the Asbestos Hazard Response Act of 1986 (H.R. 5073) being considered by this body. Several schools in Nebraska's Second Congressional District have expressed concern to me over the asbestos abatement issue, and this legislation will go a long way toward resolving problems in this area.

> The Environmental Protection Agency (EPA) estimates that some 15 million children and 1.5 million employees attend school and work in buildings contaminated by friable asbestos. Asbestos is a known human carcinogen which can cause lung cancer, mesothelioma and asbestosis — a debilitating lung disease — when airborne fibers are inhaled. Since children breathe five times faster than adults, they are much more susceptible to the effects of asbestos inhalation.
>
> In response to this problem, the EPA banned all uses of sprayed-on asbestos in 1978, and followed this action by promulgating regulations that require inspection of schools and notification of parent-teacher associations if friable asbestos is found. Additionally, EPA is also currently implementing the Asbestos School Hazard Abatement Act (ASHAA), which was enacted in 1984, in an effort to strengthen the Nation's asbestos-in-schools program by providing $100 million per year in grants and interest-free loans for school asbestos cleanup efforts.
>
> However, none of these efforts actually require asbestos abatement. The legislation we are considering today that requires EPA to provide school officials with regulations to properly inspect and abate asbestos will more fully address the problem. Moreover, the bill will also mandate the establishment of a model contractor accreditation plan so that those hired to do cleanup work are fully trained.* Since the EPA estimates that as much as 75 percent of all cleanup work has been done improperly, this contractor accreditation plan is a key provision of H.R. 5073.
>
> Finally, Mr. Speaker, the legislation we are considering today can ameliorate the problems of finding affordable insurance for asbestos work by providing a national standard for asbestos abatement. I have had input from contractors in my district outlining the problems in this area, and this legislation provides the mechanism by which affordable insurance will once again became available for asbestos work.
>
> In the aggregate, H.R. 5073 provides the framework for addressing the grave asbestos problem in schools and I urge the members to support this bill.[24]

Congress wanted asbestos properly identified and efficiently removed, but no one addressed the uncertainties associated with asbestos in school buildings, and everybody equated asbestos abatement with asbestos removal. Knowledgeable Congressional discussion of risk assessment remained nonexistent.

HR Bill 5073

The bill in the House of Representatives (HR 5073) required EPA to issue regulations within 360 days of enactment. The legislation required regulation to do the following:

* One of the serious problems associated with asbestos abatement was that "rip and skip" contractors who carelessly, and in an incompetent manner, removed asbestos, thus causing harm to those hired to remove the asbestos who were often unaware of any of the risks associated with asbestos removal. The contractors often overcharged, sometimes said asbestos existed in a building when it did not, and inadvertently raised dust levels higher than they were before rip out.

- prescribe inspection procedures to determine the presence of asbestos;
- define appropriate response action to protect against the adverse effects of asbestos;
- provide for implementation of response action;
- determine when response action is completed;
- set standards for education of workers and protection of building occupants;
- require local educational agencies to develop asbestos-management plans;
- initiate operation, maintenance, and repair for asbestos remaining in schools;
- insure safe transportation and disposal of asbestos.[25]

EPA, along with the requirements to prescribe proper inspection procedures, and to define circumstances when asbestos should be cleaned up, was required to establish a contractor accreditation program.

The bill was unanimously passed in the House of Representatives. Discussion in the Senate was remarkably similar to discussion in the House of Representatives. For example, Mr. Mitchell said, "I am very pleased that we have before us today legislation to encourage the identification and abatement of asbestos containing materials in our Nation's schools."[26] Or, Mr. Baucus, who by praising the pending legislation and informing his colleagues of the numerous groups in favor of the bill, said,

> It is well documented that when released into the air, asbestos fibers can cause lung cancer, asbestosis, pleural mesothelioma and other debilitating lung diseases.
>
> Children are particularly vulnerable to developing these diseases due to greater sensitivity, higher respiration rates, and a longer remaining lifetime during which the diseases may develop.
>
> Scientists have not demonstrated a threshold level below which exposure to asbestos can be considered safe. Many scientists believe there is no safe level of exposure to asbestos. The increase in the number of cancers linked to long-term asbestos exposure is the result of inadequate control of asbestos in previous decades. Enactment of this legislation will insure that additional avoidable deaths do not occur.[27]

Discussion of risk, reference to risk assessment, or even a discussion of the possibility that uncertainty existed did not appear in the Congressional Record. The same myths about asbestos in buildings continued to persist in the popular mind. No wonder the Asbestos Hazard Emergency Response Act was passed unanimously and signed into law in 1986.

Moral: If you repeat something enough times it becomes accepted as a "scientific" truth and imbedded as such in the public mind. EPA could have been more effective and constructive had it chosen to educate the public and the Congress. But it did not. The agency's inability to lead and educate derived from its own lack of leadership and limited understanding of the issues, both scientific and social.

AHERA solved nothing. The new law increased the very problems for which it sought solutions. The objective of AHERA was to protect school children and others who work in schools, and to solve the problem of poorly performed and unnecessary removal of ACM. It did not. It made conditions even worse, often creating health hazards for those who removed the asbestos. The AHERA mandate, to protect human health required EPA to promulgate regulations to assure minimal exposure to airborne asbestos fibers. The regulations designed to correct problems existing in 1986 did not. Before AHERA was passed, school officials lacked the needed scientific and technical information to make informed decisions about health risks associated with ACM in schools. Parents demanded responses, and school officials followed EPA guidance, applied visual and subjective criteria (ones without scientific or proven relationship to asbestos risks), and inevitably undertook expensive asbestos removal or abatement projects. The passage of AHERA might have, but did not, provide objective criteria to be utilized for assessment and for a new and rational response to the presence of ACM in schools. The need for objective, scientifically-based standards on which school officials could base decisions remained unmet.

COMMUNICATING ABOUT RISK

EPA retrospectively characterized its own role in communicating about health risks associated with asbestos in schools in a 1992 report entitled, "Communicating About Risk: EPA and Asbestos in Schools."[28] The document reviewed the role EPA communications policies and information played in asbestos management decisions. It was commissioned by EPA Administrator William K. Reilly. EPA concluded the following for the period before and leading to AHERA.

> The content analysis and anecdotal information collected through the outreach effort lead to the conclusion that EPA emphasized removal as the primary means of controlling asbestos risk.[29]
>
> EPA has inadvertently led to the confusion by issuing evolving and sometimes what may appear to be conflicting messages over time.[30]
>
> On enforcement policy the document stated,
>
> Before AHERA, there was an asbestos inspection rule requiring schools to identify asbestos in their buildings. When school compliance with this rule proved extremely low (i.e., less than 50 percent) senior EPA officials shaped up a rhetorical campaign (mainly through public speeches) emphasizing the risks of asbestos and the need for compliance with the inspection rule. EPA also began to publicize enforcement actions against schools which did not comply with the rule. These actions may have fed public perceptions that removal was the best way to avoid problems with EPA.[31]

The EPA document also noted that messages from other EPA programs conflicted with asbestos in schools messages. For example, NESHAPS-Asbestos requirements under the Clean Air Act called for removal of asbestos prior to demolition or renovation in buildings. Thus, the message received by the public was that asbestos in schools and public buildings is dangerous and should be removed. The Office of Toxic Substances also sent the message that asbestos is dangerous and needs to be removed when it banned further manufacture of asbestos-containing products in the United States.[32] EPA's messages about health risks and the agency's inability to explain and clarify the issue to the public and to Congress only made the agency less able to propose and carry out a protective, rational, public health policy for asbestos in schools. Looking backward, perhaps better understanding of the risks associated with asbestos in schools by EPA might have led the agency to create a more rational policy and allowed it to lead rather than be buffeted by the Congress and others with vested interests in the outcome. Critical strategies about how to protect, what to protect, and the cost of protection were never adopted by EPA, thus paving the way for the fiasco of AHERA.

RULEMAKING

After Congress passed AHERA, EPA prepared for rulemaking. The Act required EPA to issue proposed rules by April 20, 1987 (180 days after enactment) and to issue final rules by October 17, 1987. The process of developing regulations, called negotiated rulemaking, was carried out by groups with interests affected by the regulations. A committee was created to develop proposed rules. The rulemakers, established as a Federal Advisory Committee, consisted of representatives of national education organizations, labor unions, asbestos product manufacturers, the environmental community, asbestos abatement contractors, professional associations of architects, consulting engineers, industrial hygienists, states, and EPA. The members of the Advisory Committee represented 23 interests in all.[33] The Committee met for 11 days, from January to April 1987. The National School Boards Association, the Environmental Defense Fund, the National PTA, the Association of Wall and Ceiling Industries, manufacturers of asbestos surfacing, pipe, and block insulation, and the Association of State Attorneys General, as well as EPA staff, attended the meetings. The proposed rule agreed to by most members (20 of 24) was published on April 30, 1987.[34] It included rules for inspection requirements, sampling, assessment, training, management, labeling, response actions, accreditation, and enforcement.

Criticism of the proposed rules quickly followed their publication. Comments by Michael Gough on the proposed rule sums up the criticisms.

1. The proposed regulation requires that likelihood of exposure to asbestos be estimated by visual inspection of asbestos-containing material in schools. Decisions whether to remove asbestos, to abate possible exposures through measures other than removal or to manage asbestos in place were to be based on the visual inspections. But efforts by the EPA and others to demonstrate correlation between visual inspection results and airborne asbestos have failed.
2. The only way to determine levels of airborne asbestos in school buildings is to use air monitoring techniques that quantify the amount of asbestos fibers in air using an electron microscope. EPA does not require or consider air monitoring to assess possible exposure. Thus, failure to monitor air makes it impossible to assess the possible risk in a school before making a decision about what action to take. The proposed regulation

discourages air monitoring and prevents exposure assessment, an essential component of any sound risk assessment.

3. The rulemaking assessment used estimates of asbestos-in-air that were far higher than actual measurements made by other investigators. Thus, the rulemaking risk assessment "cannot be accepted as representative of risks in the nation's schools."
4. The proposed rules lack appropriate risk assessment criteria and may be harmful. Proposed allowable levels of exposure after remediation are higher than levels reported in most buildings that have been examined.
5. There is no other valid way other than air monitoring to determine exposure levels for the purpose of risk assessment. Without air monitoring a number of undesirable outcomes will follow. For example, significant funds will be spent with little or no gain in health protection, positive harm to health could result.[35]

According to Gough and a growing list of scientists, air monitoring (conspicuously absent from the rules) offered the only rational guide to decisions that will protect human health by the least burdensome means.[36]

EPA, consistent with its earlier approach to asbestos in schools, ignored the role of air sampling in the assessment of health risk from asbestos in schools. The Congress had charged the agency with providing objective standards to trigger response. Instead, EPA proposed recourse to "experts" to judge the ACM using inadequately defined terms in the proposed rule. And the experts seldom agreed. Some experts believed in removal of ACM in almost all cases. Others believed that the risks of any removal are great because they elevate airborne concentrations of asbestos. The proposed rules were based on the premise that qualitative assessment by experts could characterize the risks (present and future) from inhalation of asbestos-in-air in the environment.

But air sampling is the major tool for determining risk from inhaled toxic materials, and EPA took the position in the case of ACM in schools that it would not rely on measurement of toxic materials (asbestos) in air.[37] The agency continued to promote confusion rather than clarify the issues. It continued to promote estimated risks based on incorrect qualitative assumptions rather than quantitative measurements. The near panic response to asbestos initiated earlier by EPA and based on its overestimation of the magnitude of risk, instead of slowing down after AHERA and AHERA rulemaking, speeded up.

REFERENCES

1. Ontario Ministry of the Attorney General, *Report of the Royal Commission on Matters of Health and Safety Arising from the Use of Asbestos in Ontario,* Ontario, Toronto, 1984.
2. Ontario Ministry of the Attorney General, *Report of the Royal Commission on Matters of Health and Safety Arising from the Use of Asbestos in Ontario,* Vol. II, Ontario, Toronto, 1984, 548.

3. House of Representatives, 99th Congress, *Hearings Before the Subcommittee on Commerce, Transportation and Tourism*, EPA Efforts to Control Asbestos Hazards, Serial Number 99-77, First Session, June 27, 1985, United States Government Printing Office, Washington, D.C., 1986.
4. House of Representatives, 99th Congress, *Hearings Before the Subcommittee on Commerce, Transportation and Tourism*, EPA Efforts to Control Asbestos Hazards, Serial Number 99-77, First Session, June 27, 1985, United States Government Printing Office, Washington, D.C., 1986, II.
5. House of Representatives, 99th Congress, *Hearings Before the Subcommittee on Commerce, Transportation and Tourism*, EPA Efforts to Control Asbestos Hazards, Serial Number 99-77, First Session, June 27, 1985, United States Government Printing Office, Washington, D.C., 1986, III.
6. House of Representatives, 99th Congress, *Hearings Before the Subcommittee on Commerce, Transportation and Tourism*, EPA Efforts to Control Asbestos Hazards, Serial Number 99-77, First Session, June 27, 1985, United States Government Printing Office, Washington, D.C., 1986, 1.
7. House of Representative, 99th Congress, *Hearings Before the Subcommittee on Commerce, Transportation and Tourism*, EPA Efforts to Control Asbestos Hazards, Serial Number 99-77, First Session, June 27, 1985, United States Government Printing Office, Washington, D.C., 1986, 2.
8. House of Representatives, 99th Congress, *Hearings Before the Subcommittee on Commerce, Transportation and Tourism*, EPA Efforts to Control Asbestos Hazards, Serial Number 99-77, First Session, June 27, 1985, United States Government Printing Office, Washington, D.C., 1986, 3.
9. House of Representatives, 99th Congress, *Hearings Before the Subcommittee on Commerce, Transportation and Tourism*, EPA Efforts to Control Asbestos Hazards, Serial Number 99-77, First Session, June 27, 1985, United States Government Printing Office, Washington, D.C., 1986, 10 (Statement of John A. Moore).
10. House of Representatives, *Hearings*, 99th Congress, *Hearings Before the Subcommittee on Commerce, Transportation and Tourism*, EPA Efforts to Control Asbestos Hazards, Serial Number 99-77, First Session, June 27, 1985, United States Government Printing Office, Washington, D.C., 1986, 12 (Statement of John A. Moore).
11. House of Representatives, 99th Congress, *Hearings Before the Subcommittee on Commerce, Transportation and Tourism*, EPA Efforts to Control Asbestos Hazards, Serial Number 99-77, First Session, June 27, 1985, United States Government Printing Office, Washington, D.C., 1986 35.
12. House of Representatives, 99th Congress, *Hearings Before the Subcommittee on Commerce, Transportation and Tourism*, EPA Efforts to Control Asbestos Hazards, Serial Number 99-77, First Session, June 27, 1985, United States Government Printing Office, Washington, D.C., 1986, 37.
13. House of Representatives, 99th Congress, *Hearings Before the Subcommittee on Commerce, Transportation and Tourism*, EPA Efforts to Control Asbestos Hazards, Serial Number 99-77, First Session, June 27, 1985, United States Government Printing Office, Washington, D.C., 1986, 39.
14. House of Representatives, 99th Congress, *Hearings Before the Subcommittee on Commerce, Transportation and Tourism*, EPA Efforts to Control Asbestos Hazards, Serial Number 99-77, First Session, June 27, 1985, United States Government Printing Office, Washington, D.C., 1986, 52.

15. House of Representative, 99th Congress, *Hearings Before the Subcommittee on Commerce, Transportation and Tourism*, EPA Efforts to Control Asbestos Hazards, Serial Number 99-77, First Session, June 27, 1985, United States Government Printing Office, Washington, D.C., 1986, 56.
16. House of Representatives, 99th Congress, *Hearings Before the Subcommittee on Commerce, Transportation and Tourism*, EPA Efforts to Control Asbestos Hazards, Serial Number 99-77, First Session, June 27, 1985, United States Government Printing Office, Washington, D.C., 1986, 145, (EPA Memorandum from Alvin Alm, Deputy Administrator to Regional Administrators (Regions 1-10) August 1984.
17. House of Representatives, 99th Congress, *Hearings Before the Subcommittee on Commerce, Transportation and Tourism*, EPA Efforts to Control Asbestos Hazards, Serial Number 99-77, First Session, June 27, 1985, United States Government Printing Office, Washington, D.C., 1986, 161.
18. House of Representatives, 99th Congress, *Hearings Before the Subcommittee on Commerce, Transportation and Tourism*, EPA Efforts to Control Asbestos Hazards, Serial Number 99-77, First Session, June 27, 1985, United States Government Printing Office, Washington, D.C., 1986, 179.
19. House of Representatives, 99th Congress, *Hearings Before the Subcommittee on Commerce, Transportation and Tourism*, EPA Efforts to Control Asbestos Hazards, Serial Number 99-77, First Session, June 27, 1985, United States Government Printing Office, Washington, D.C., 1986, 190.
20. House of Representatives, 99th Congress, *Hearings Before the Subcommittee on Commerce, Transportation and Tourism*, EPA Efforts to Control Asbestos Hazards, Serial Number 99-77, First Session, June 27, 1985, United States Government Printing Office, Washington, D.C., 1986, 201.
21. House of Representative, 99th Congress, *Hearings Before the Subcommittee on Commerce, Transportation and Tourism*, EPA Efforts to Control Asbestos Hazards, Serial Number 99-77, First Session, June 27, 1985, United States Government Printing Office, Washington, D.C., 1986, 193.
22. House of Representatives, 99th Congress, *Hearings Before the Subcommittee on Commerce Transportation and Tourism,* Asbestos Hazard Emergency Response Act of 1986, Serial Number 99-77, Second Session, March 4, 1986, United States Government Printing Office, Washington, D.C., 1986, 280.
23. House of Representatives, 99th Congress, *Hearings Before the Subcommittee on Commerce Transportation and Tourism,* Asbestos Hazard Emergency Response Act of 1986, Serial Number 99-77, Second Session, March 4, 1986, United States Government Printing Office, Washington, D.C., 1986, 222.
24. *Congressional Record*, 99th Congress, 2nd Session, Aug. 12, 1996, H5997.
25. *Congressional Record*, 99th Congress, 2nd Session, Aug. 12, 1996, H5998.
26. *Congressional Record*, 99th Congress, 2nd Session, Sept. 10, 1986, S12336.
27. *Congressional Record*, 99th Congress, 2nd Session, Sept. 10, 1986, S12337.
28. USEPA, *Communicating About Risk: EPA and Asbestos in Schools.* Final Report of the Internal Task Force, Internal Document. Jan. 1992.
29. USEPA, *Communicating About Risk: EPA and Asbestos in Schools.* Final Report of the Internal Task Force, Internal Document, Jan. 1992, 12.
30. USEPA, *Communicating About Risk: EPA and Asbestos in Schools.* Final Report of the Internal Task Force, Internal Document, Jan. 1992, 29.
31. USEPA, *Communicating About Risk: EPA and Asbestos in Schools.* Final Report of the Internal Task Force, Internal Document, Jan. 1992, 30.

32. USEPA, *Communicating About Risk: EPA and Asbestos in Schools*. Final Report of the Internal Task Force, Internal Document, Jan. 1992, 31.
33. *Federal Register*, Asbestos containing materials in schools; proposed rule and model accreditation pan; rule Vol. 52, #82, April 30, 1987, 15821.
34. *Federal Register*, Asbestos containing materials in schools; proposed rule and model accreditation pan; rule Vol. 52, #82, April 30, 1987, 15833.
35. Gough, M., Comments on EPA's proposed asbestos in schools rule, unpublished.
36. Gough, M., Comments on EPA's proposed asbestos in schools rule, unpublished, 1.
37a. Paul, J.M., Corn, M., Lees, P.S.J., and Breysse, P., Non-occupational exposure to asbestos in buildings: a practical risk management program, *Am. Indust. Hygiene Assoc. J.*, 47, 497, 1986.
37b. Corn, M., Asbestos and disease: an industrial hygienist's perspective, *Am. Indust. Hygiene Assoc. J.*, 47, 515, 1986.

Unintended Consequences

6 After AHERA

RULEMAKING

When Congress directed the Environmental Protection Agency (EPA) under the Asbestos Hazard Emergency Response Act (AHERA) to set required response actions for schools, "using the least burdensome methods which protect human health and the environment,"[1] they intended that EPA present available options that would allow for intelligent and prudent action to protect public health. However, EPA's final rule, required by AHERA, did little to alleviate previous problems. It caused more problems and did not solve the already existing ones. Rulemaking did not provide the guidance necessary for the level of risk that would dictate when and what action should be taken. Even after enactment of AHERA, which required the agency to regulate response actions to asbestos-containing materials (ACM) found in schools, EPA's approach remained inconsistent and continued to cause more public confusion and engender fear. This perpetuated poorly-planned, rapidly-executed, expensive remedial actions, the very kinds of approach and actions Congress intended to end with AHERA.

AHERA required EPA to "provide for establishment of Federal regulations (rulemaking) which require implementation of appropriate response actions with respect to asbestos-containing material in the nation's schools in a safe and complete manner."[2] It also required EPA to, "define the appropriate response actions" to carry out the mandate.[3] This requirement made initial assessment of a school building's condition crucial to the provision of essential information to the decision-making process necessary to take correct action regarding ACM in a building. But, consistent with the agency's pre-AHERA approach to ACM, EPA refused to recognize the role of air sampling in initial assessments of asbestos in buildings. The Congress charged EPA to provide objective standards which would trigger response actions. Instead, EPA responded by proposing recourse to "experts" who would "judge" the ACM by utilizing poorly defined terms cited in the Proposed Rules. For example, these terms included: damaged ACM, significantly damaged ACM, potential for future damage, and poor cohesion and adhesion. None of these terms are subject to measurement. The rule also contained numerous terms "experts" did not agree on. Some experts had already removed ACM in 99% of places where it was present. Other experts believed the risks associated with any removal job are greater in terms of exposure during removal, and of elevated airborne asbestos fiber concentrations after removal. Those experts seldom recommended removal. EPA, charged by Congress to develop appropriate responses, i.e., a defensible set of objective standards, and a decision tree for removal or in-place management, did not do so. Based on investigations by its own contractors, and confirmed by other investigations, the Agency knew that qualitative factors associated with the school environment could not be correlated with asbestos-in-air concentrations.[4] Nevertheless, the foundation of EPA's proposed rules was the premise that qualitative assessment by "experts" could properly characterize the present and future risks from inhalation of asbestos-in-air in the school environment.

AIR SAMPLING

Air sampling was, and still remains, the major tool utilized to determine the risk from inhaled toxic materials.[5] Strangely, the same agency that stressed measuring pollutants in its air pollution, water pollution, ionizing radiation, and hazardous waste programs took a position, in the case of ACM in schools, that did not rely on measurement of asbestos-in-air. In 1986, air sampling was considered a science, with numerous practitioners working in the United States and other countries. It is difficult to understand why EPA did not establish procedures that would lead to the consistent use of this scientific tool to establish a decision framework for various forms of action to be taken after determination of asbestos-in-air concentrations in a building containing asbestos. Even stranger, the Agency invoked the use of a clearance level for asbestos, which utilized air sampling for asbestos-in-air after removal, while it neglected and did not require or recommend air sampling prior to removal.

Early in its asbestos program, EPA had initiated a near panic response to asbestos in schools when they estimated risks based on incorrect assumptions of occupant exposure. Those estimated risks drove the entire asbestos regulatory program. In 1987, air sampling during initial assessment of a building could have offered an effective basis for a rational approach to deal with asbestos in school buildings, as the new law intended. Instead, EPA once again rejected air sampling for initial assessment only to later defend air sampling for clearance purposes.[6] Such a policy is reminiscent of *Alice in Wonderland* when the Red Queen insists, "Verdict first trial after."

In the Proposed Rule, EPA referred to its reluctance to press for such selected highly controversial issues during rulemaking. EPA said it tried during rulemaking to placate parties who held different views on ACM in schools and developed a resulting weak rule. More *Alice in Wonderland.* Unfortunately, the weak rule failed to offer guidance to regulatees, as Congress directed. By striking a compromise on given issues, because the issues were controversial, EPA neglected its responsibility as stated in AHERA.

Perhaps the most irresponsible abrogation of responsibility during rulemaking was EPA's failure to issue a risk assessment for asbestos-in-air of school buildings. A risk assessment, utilizing scientific evidence, would have of necessity, been based on airborne concentrations of asbestos in buildings and on exposure assessment for occupants. The integration of air sampling with the presentation of a risk assessment could have solved the burden assigned to EPA by Congress, namely, to develop specific guidelines for action keyed to the level of risk. In 1986, there were data on airborne concentrations of asbestos in the air of schools, including a Health Hazard

Assessment Document for Asbestos, a product of EPA's own research and development office.[7] Other risk assessments existed in the scientific literature.[8] EPA rulemaking, which was meant to provide guidance, instead made the issue more confusing and offered no guidance because the rules were not concrete even though Congress had enacted AHERA to correct that fundamental problem. The regulatory agency did nothing to define which response actions would protect health and which would be least burdensome. The role of a regulatory agency to determine if exposure levels to asbestos in schools were safe and what response actions would make them safer was ignored by EPA. It opposed any requirement to adopt uniform national standards governing asbestos responses and insisted that setting standards would result in schools losing the flexibility needed to make appropriate decisions. Incomprehensively, according to an Agency spokesman, loss of flexibility would lead to, "unnecessary and potentially dangerous abatement activity."[9]

What level of asbestos exposure in schools is safe? Will removal make schools safer? These key questions went unanswered. EPA did not determine what level of asbestos is safe. It did not provide criteria to distinguish safe schools from unsafe schools. Furthermore, EPA did not demonstrate that removal of asbestos made schools safer than regular maintenance, referred to as O & M, Operations and Management.

FINAL RULE

The final rule required local education agencies (LEAs) to have inspections conducted by an accredited inspector who would provide a written assessment. The LEA must then select a person accredited to develop management plans, review results of inspections, and recommend appropriate response actions. Then the LEAs must select and implement, in a timely manner, the appropriate response actions consistent with the assessment.[10] Rulemaking also developed model accreditation regulations. The regulations specified separate accreditation requirements for inspectors, management planners, and those who design and carry out response actions.

> Persons in each of the above disciplines perform a different function. Inspectors identify and access the ACM's condition. Management planners use the data gathered by inspectors to assess the ACM's hazard, determine the appropriate response actions and develop a schedule for implementing response actions. Abatement project designers determine how the asbestos abatement work should be conducted. Lastly, asbestos abatement contractors, supervisors and workers carry out the abatement work."[11]

The length of courses required for the above vary. The rules required inspectors to take a three-day course. Management planners must take the inspection course plus two more days devoted to management planning. Abatement project designers required at least three days of training. Abatement contractors and supervisors required a four-day training course, and abatement workers required a three-day training course.[12] The men and women who attended and passed these accredited courses were often referred to as three-day wonders. Responsibility for accreditation of responsible persons was assigned to individual states, thus, introducing enormous variations in the quality of those accredited.

ASBESTOS ABATEMENT INDUSTRY

Even before AHERA, an asbestos abatement industry had begun to develop in response to earlier laws. After AHERA, the growth of the asbestos abatement industry was unprecedented. The industry began in 1979 when EPA established the voluntary Technical Assistance Program (TAP) to encourage school administrators to test for and abate asbestos in their buildings. After a very limited response on the part of school administrators to TAP, EPA established the Asbestos-in-Schools Identification and Notification Rule (ASINR). It required all public and private secondary schools to inspect for ACM, analyze samples for asbestos content, and notify all school employees and parent-teacher groups or parents as to their findings. EPA itself determined that less than 10% of the schools were in full compliance with ASINR and some schools, although in compliance, had trouble financing their asbestos abatement projects. In response, Congress enacted the Asbestos School Hazard Abatement Act of 1984 to enable schools to obtain loans and grants to pay for asbestos inspection and abatement. All of this activity encouraged asbestos abatement contractors and stimulated the rapid growth of the asbestos abatement industry. Many of the asbestos abatement contractors entered the field ill-educated and poorly-equipped to undertake the job of asbestos abatement safely and competently. As a result, the asbestos hazard worsened in many schools.[13] Workers hired to perform abatement jobs were too often exposed to high levels of asbestos fibers as they ripped out asbestos from walls and ceilings and created dangerous dust. Often they remained unaware of the hazards associated with asbestos at their job.[14]

To collect the cost of removal, school districts throughout the United States sued for property damage. The United States District Court for Eastern Pennsylvania certified a nationwide class action on behalf of all school districts against companies which made and distributed asbestos products in order to recover costs associated with asbestos removal.[15] Many individual schools retained the option of not participating in the nationwide class action. Larger school districts filed their own actions. Asbestos in buildings litigation was fast becoming a growth industry. The growing trend to remove asbestos from schools and the associated legal activity directly resulted from EPA's prediction in 1982 that up to 40,000 school children a year would die from cancer because the schools they attended contained asbestos. It was a prediction based on two false premises. First, the presence of any asbestos in a building is dangerous no matter what the amount or how much is breathed. The idea that "one fiber can kill" had become extremely popular and imbedded in the public mind, and although unproven led to the common perception that asbestos, no matter how small the amount, killed. Second, it was assumed that in schools and public buildings non-occupational risks could be equated with epidemiological studies of workers exposed to high levels of asbestos concentrations at work (occupational) many years earlier. Based on these premises, EPA had encouraged unnecessary removal and generated fear of adverse health risk which bordered on hysteria.

Cost of Removal

The economic costs of asbestos removal continued to rise, unchecked by AHERA. An imbalance of perception of hazard and allocation of resources surely existed. Estimates of cost vary. In the proposed AHERA rules (1987), EPA had projected the economic impact of AHERA by estimating the incremental costs attributable to the proposed regulations. The analysis included costs of inspection, sampling, development and management plans, implementation of response actions, periodic surveillance, and provision of required training. "The estimated net present value of the costs of these proposed regulations is approximately $3,219 million (using a 10% discount rate) over 30 years. This includes the cost of initial inspection and sampling — $58.2 million; development and implementation of management plans — $970.8 million; periodic surveillance — $41.8 million; reinspection — $34.7 million; special operations and maintenance programs — $525.4 million; and abatement response actions — $1,587.8 million."[16]

In 1989, according to Kate Herber, legislative council for the National School Board Association, costs of asbestos abatement for elementary and secondary schools could be as high as $6 billion. The figure reflects removal costs for 671 of the more than 15,000 public school districts around the country who answered a survey. The cost of compliance just for surveyed schools from fiscal 1986 to 1990 would approach $464 million.[17] Obviously, a large discrepancy existed between agency estimated costs and costs actually incurred by school districts.

The asbestos removal industry also estimated costs. Their estimates included removal from commercial as well as school buildings and described a continued growth rate in 1990. A market of $8 billion in the mid-1990s is not unreasonable, and was reported to attendees at the 7th Annual Asbestos Abatement Conference and Exposition sponsored by the National Asbestos Council. Estimates given at the meeting were based on information gathered by the National Insulation and Abatement Contractors Association from 3,500 companies. The National Asbestos Council estimated a total market between $4.2 and $4.4 billion; figures well above the 1988 market of $3 billion.[18] A major factor during the market was federal, state, and local regulations. Spending on asbestos abatement for schools and commercial buildings from 1986 to 1992 is shown graphically in Table 1.[19]

Public concern, Congressional action, and EPA regulations initially stimulated the growth of the asbestos removal industry. The well-organized, self-interested asbestos removal industry, once started, drove asbestos removal. Removal triggered lawsuits against suppliers and manufacturers of ACM for removal costs and damages to occupant's health. Plaintiffs often sought punitive damages for failure to warn of health risks associated with ACM in buildings. Court cases continue to this day. Removals continue despite a developing literature contradicting EPA's "no threshold" dogma for cancer induction and no demonstration that asbestos levels in schools are carcinogenic. Table 2 lists school system suits.

Table 2 displays the number of cases filed on behalf of schools in the United States. Many individual school buildings and/or school systems can be part of a single case. Thus, Table 2 data do not permit one to estimate the number of school buildings involved in litigation. Cases were filed against other suppliers in addition

TABLE 1
Spending in U.S. on Asbestos Abatement (in Billions)

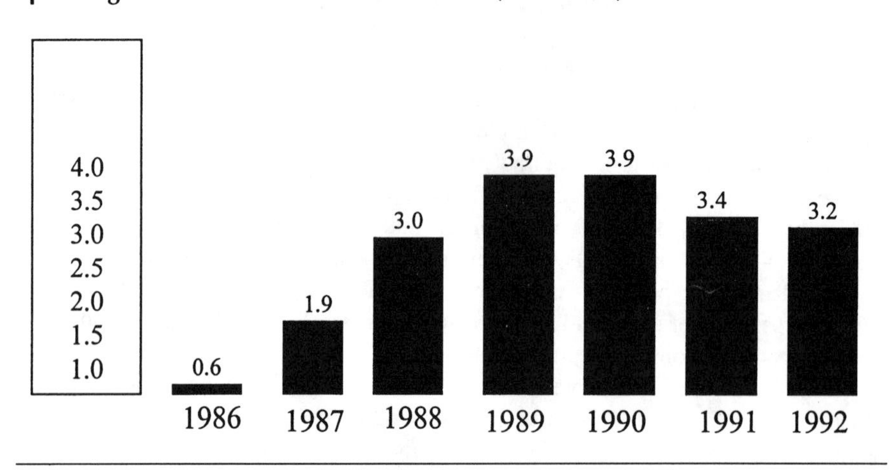

to the three suppliers associated with Table 2. Information on the other suppliers could not be obtained.

NEW YORK CITY

Misunderstanding, manipulation of science, mismanagement, greed, ineptitude, and politics as usual, stimulated by a federal law, had created an asbestos crisis. The response to asbestos in schools in New York City represents a worst case scenario. Problems surfaced when the New York City Board of Education postponed opening its 1,096 schools in the fall of 1993 to complete an emergency asbestos inspection of all New York City Schools. The debacle that followed, the great New York City asbestos panic, occurred even after many studies had already appeared challenging the conventional wisdom about asbestos. A number of scientists publishing their research results in reputable scientific journals had already concluded that the risk of disease from exposure to ACM in buildings had more to do with hysteria than reality.*[21] The problems previously discussed in this chapter and related to EPA rules for AHERA contributed to the New York City's school asbestos crisis.

In August 1993, the Board of Education directed the New York City School Construction Authority to carry out a crash reinspection of all New York City schools within seven weeks. The directive to inspect all 1,096 schools delayed the opening of schools from the previously scheduled opening day, September 2, 1993 to September

* Some of the scientists who concluded that risks of disease from exposure to asbestos-containing materials in buildings were minimal included: Hans Weill and Janet Hughes, Tulane University; Richard Doll and Julian Peto, Oxford University; Brook Mossman, University of Vermont; Morton Corn, Johns Hopkins University. *The Report of the Royal Commission, Matters of Health and Safety Arising from the Use of Asbestos in Ontario*, reached the same conclusion.

TABLE 2
Cases Filed by Schools Against Three Former Asbestos Suppliers of Insulation and Ceiling Products, 1982–1995

State in which case was filed	Number of cases filed	State in which case was filed	Number of cases filed
Alabama	3	Nevada	
Alaska	2	New Hampshire	2
Arizona	1	New Jersey	6
Arkansas		New Mexico	
California	3	New York	3
Colorado	4	North Carolina	4
Connecticut	2	North Dakota	8
Delaware		Ohio	16
Florida	6	Oklahoma	
Georgia	3	Oregon	2
Hawaii	2	Pennsylvania	10
Idaho		Rhode Island	2
Illinois	3	South Carolina	21
Indiana	1	South Dakota	6
Iowa		Tennessee	17
Kansas	1	Texas	9
Kentucky	10	Utah	
Louisiana		Vermont	1
Maine	4	Virginia	5
Maryland	4	Washington	3
Massachusetts	2	West Virginia	1
Michigan	4	Wisconsin	9
Minnesota	4	Wyoming	1
Mississippi	6	District of Columbia	1
Missouri	17	Location Uncertain	94
Montana		University Cases	19
Nebraska	8	Total Case Count	330

20, 1993. Over 1,000,000 children attend New York City's elementary, junior high, and high schools. The reason given for the closings was possible asbestos exposure.

This fiasco can be traced back to 1979 (14 years earlier) when the citizens of New York City, frightened by newspaper and other media coverage of asbestos, and believing that ACM in schools threatened their children's health, heeded EPA's warnings and demanded removal of ACM from city schools. Nobody defined safety.

Nevertheless, parents sought absolutely safe schools. In 1979, the New York State Legislature passed one of the first state asbestos laws. It called on local school districts to survey buildings, identify asbestos, and clean up hazards. The New York City Board of Education formed an Asbestos Task Force and installed it in the Division of School Building offices. The Task Force was nicknamed the flying squad, because its members rushed to schools to clean up asbestos. 1979 was the same year

EPA established TAP, a federal voluntary technical assistance program to encourage school administrators to test for and abate asbestos in their school buildings. In 1979, EPA also published its first guidance document, Asbestos Containing Materials in School Buildings.[22] It is clear that school systems throughout the United States, not just New York City, required guidance. They had similar problems, and did not know how to solve them. What they all had in common was fear of asbestos.

Limited response to TAP led EPA to establish the Asbestos-In-Schools Identification and Notification Rule (ASINR). The rule required inspection for asbestos-containing friable materials, analysis of samples for asbestos content, and notification of parent/teacher groups and all employees of the findings. EPA later determined that less than 10% of schools in the United States were in compliance and those in compliance had trouble financing asbestos abatement.[24] A series of other poorly designed rules, and numerous EPA guidance documents further complicated the federal role in asbestos abatement. The New York City Asbestos Task Force had a history of incompetence managing state mandated and federally mandated inspections.

In 1986, after Congress passed AHERA, all the nation's school districts were required to appoint a local official to oversee the inventory of asbestos in school buildings and to develop plans to eliminate the hazard. The Board of Education in New York City chose an architect, Robert Pardi, from their division of School Buildings to head the Asbestos Task Force. He had no background in asbestos, and limited administrative experience. In 1988, Pardi hired 26, mostly foreign born, inspectors. Many were previously unemployed. Some had engineering backgrounds. None had previous asbestos experience. They started work at an approximate salary of $26,000.[24] In 1986, before the Board of Education hired Pardi, it closed a school for handicapped children in the Washington Heights section of New York City after the Task Force had failed to adequately supervise an asbestos removal job. In 1987, the Asbestos Task Force concluded that all testing it performed in the seven years previous to 1987 had to be redone. Mismanagement was just part of the Task Force history.

In 1988, New York City Board of Education officials hired the Envirosafe Corporation, later renamed Envirospan Safety Corporation and signed a $1.4 million contract for Envirosafe to design the asbestos survey and to train and supervise asbestos inspectors. Envirosafe had no experience in conducting large and complex inspections. The training took place at Envirosafe headquarters in Manhattan. Training included three days on sampling techniques and two days on how to draw up plans for managing hazardous asbestos found in schools. EPA approved the course. Inspectors interviewed in 1993 said they did not learn or have a sampling methodology, the number of samples required was too low, and after sampling they (the inspectors) could not pinpoint the location of samples they took. They complained about lack of supervision and staff turnover.[25] In 1989, a frantic group of inspectors rushed to meet a May deadline. Crews worked 14-hour days and earned large amounts of overtime. The Task Force, according to a New York Times report, renamed the Environmental Health and Safety Group, would become the major focus of an investigation into mishandling and possibly fraudulent asbestos testing.[26] Inexperienced workers raced around to complete the job of finding asbestos and bungled the job. Fraudulent reports seemed to be the order of the day. Confusion ruled.

When inspections in the summer of 1993 turned up buildings with asbestos that had been reported "clean" in the earlier survey, Mayor Dinkins ordered all city schools reinspected immediately. In August, investigators sealed the task force's offices and announced that hundreds of inspection reports had been falsified and thousands of documents were missing. The investigators worked for the independent School Construction Authority and the Special Commissioner for investigations of the New York City School District. "The level of the Asbestos Task Force's incompetence and management in its inspection program was astounding," said Thomas D. Thacher, Inspector General for the School Construction Authority. "They worked for seven years carrying out asbestos inspections, and then in 1987 threw out what they had already done as unreliable. So it was no surprise that it repeated the same mistakes for the next three years. That's the tragic legacy they left us today."

The Special Commissioner barred members of the Task Force from doing any further work on August 6th and locked them out of their offices. The Task Force was the major focus of the investigation. Since August 6th, the 53 members of the Asbestos Task Force have been reassigned to temporary duty unrelated to asbestos and barred by supervisors and their union from speaking publicly. But some Task Force members when promised anonymity told a reporter that their efforts were marred by staff discord, malfunctioning equipment, and confusion about how to perform inspections. One inspector who helped to survey some seventy schools said, "At that time, nobody knew what they were doing. We had to learn everything as we went along."[27]

When Mayor Dinkins closed the 1,069 New York City Public Schools and ordered immediate emergency inspection of them all, the delayed opening disrupted the families of 1,000,000 children, at a cost of tens if not hundreds of millions of dollars. A new opening day, September 20th was set. The New York City School system was in disarray. On the positive side, the asbestos crisis did point out the poor maintenance of New York City schools, the need for more money for basic school maintenance, and the wide physical deterioration of the schools.

The newly appointed Chancellor of the New York City school system, Cortines, had an overwhelming job. He and the mayor promised to inspect and open the schools by September 20th. By September 14th, it became clear that the School Construction Authority could not provide Cortines with the report he requestioned to determine which classrooms could open and which would have to remain closed. The Chancellor ordered new committees set up to prepare emergency, room by room reports. The situation was chaotic. The inaccurate School Construction Authority report gave some schools a clean bill of health. But the same schools listed in the report had been listed in other reports as needing major asbestos clean-up. Overriding the School Construction Authority Cortines ordered the establishment of a four-person committee for each school. The committee included the principal, a Teacher's Union representative, the school custodian, and a parents' association president. They would tour their building, review rooms with damaged walls and ceilings, and report to the district superintendent by 2:00 p.m. on September 14th. A memo from Cortines cited a letter from Acting Commissioner of Health, Benjamin K. Chu, about certifying which rooms are safe for occupancy. It said, "On this basis, rooms and areas that do not contain plaster damage, regardless of the fact that the building may not have undergone final Operation Clean House clearance testing, will be considered safe for occupancy on September 20."[28] On September 15th, Cortines said the schools would open on September 20th although the surveys done the day before unexpectedly closed off hundreds of additional rooms. The city's 39 district superintendents had some information to enable last minute planning for opening day. On the weekend before opening day superintendents still did not have a complete list of schools that would be closed. The status of individual schools was not yet made public. Sunday papers presented information for parents and students but the message was confused and contradictory. Information included which schools would shorten the school day, which had double sessions, and new bus schedules. Most schools opened on September 20th but it was a tumultuous day. Some buildings stayed closed, and thousands of students remained locked out. 100 school buildings stayed closed and 90 schools functioned on reduced or rearranged schedules.

Nobody asked if the children in New York City schools faced real danger from asbestos or if the asbestos threat was so great as to warrant the disruption of the schools. No mention appeared in the media or reports of the city officials about the concentrations of fibers in the air. Indeed, the law never required the Board of Education to measure asbestos fiber concentrations in air. New York City's Board of Education followed the earlier advice given by EPA which discouraged air sampling prior to abatement, despite a new 1990 EPA advisory, *The Green Book,*[29] which toned down the EPA stand and recommended operation and maintenance rather than abatement of asbestos. EPA remained silent through the crisis.

Ironically, three years prior to the frantic, hysterical occurrences in New York City, *Science* published an article entitled, "Asbestos: Scientific Developments and Implications for Public Policy."[30] The article became a catalyst in the dispute over asbestos in school buildings and EPA's policy of asbestos removal. The dispute had been developing since before AHERA's passage. It may have appeared to the public that no debate about asbestos in building existed but a controversy had been active in the scientific community around the question of in-place asbestos. It does not seem to have led to any debate during New York City asbestos crisis in 1993. According to the authors of the *Science* article, available data did not support the idea that low levels of exposure to asbestos pose a health hazard to building and school occupants. They said that the health hazard of in-place chrysotile asbestos was grossly overstated and that ripping out asbestos to solve the problem of asbestos in school buildings is probably more harmful than the problem itself. The authors also distinguished between types of asbestos. Not all types, they said, were equally unsafe. For example, chrysotile, the asbestos utilized for most building materials is not as dangerous as amphibole.

> The AHERA ruling of 1986 brought asbestos to the attention of the U.S. public and instilled fears in parents that their children would contract asbestos-related malignancies because of high levels of airborne asbestos fibers in schools. Panic has been fueled by unsupported concepts such as the "one fiber theory," which maintains that one fiber of inhaled asbestos will cause cancer. As a result of public pressure asbestos is often removed haphazardly from schools and public buildings even though most damaged ACM is in boiler rooms and other areas which are inaccessible to students or residents

> ... Does airborne asbestos present a risk to the health of individuals in schools and other buildings? The available data does not indicate that asbestos-associated malignancies or functional impairment will occur as a result of exposure to most airborne concentrations of asbestos in buildings.[31]

This article was published in January. In March, a scathing editorial appeared in *Science* entitled, "The Asbestos Removal Fiasco." The first paragraph, an opening salvo, took issue with asbestos policy and activity.

> Removal of asbestos from buildings could cost as much as $50 to $150 billion. The content of asbestos fibers in the air of buildings containing asbestos is harmlessly small and essentially the same as in outdoor air. Asbestos in buildings, unless damaged, does not shed fibers. The removal process releases asbestos fibers which could result in more cancer in the workmen than would have resulted in the usual occupants had the asbestos been left in place.[32]

The *Science* editor, Philip Abelson, noted the difference between chrysotile, a serpentine mineral and the dangerous amphibole crocidilite, puzzled by federal legislation and regulations that lump both together. He wrote that EPA fostered the view that a single fiber can cause cancer, noting that the hypothesis is unproven. The author said he is also puzzled by EPA's lack of expeditious effort to obtain rigorous measures of indoor and outdoor levels of fibers. "One would think that in a $50-to-$150 billion program the first priority would be an accurate assessment of the problem."[33] EPA credibility has been damaged, said Abelson.

Perhaps the new research would have the effect of reevaluating asbestos in school policy. The New York City experience belies that thought. But change occurs slowly. In 1990, EPA issued its fourth and last guidance document, *Managing Asbestos in Place*.[34] This new guide offered guidance for operations and maintenance (O&M) programs for asbestos in-place. But it went even further than placing importance on O&M rather than abatement. The agency conceded on five important facts. They would in the future be referred to as The Five Facts. "Fact 1: Although asbestos is hazardous, the asbestos-related disease depends on exposure to airborne asbestos fibers; Fact 2: Based upon available data, the average airborne asbestos levels in buildings seem to be very low. Accordingly, the health risk to most building occupants also appears to be very low; Fact 3: Removal is often not a building owners best course of action to reduce asbestos exposure. In fact, an improper removal can create a dangerous situation where non previously existed; Fact 4: EPA only requires asbestos removal in order to prevent significant public exposure to airborne asbestos fibers during building demolition or renovation activities; and Fact 5: EPA does recommend a pro-active, in place management program whenever asbestos-containing materials is discovered."[35]

In retrospect, EPA did not present the American public or its legislators with clear technical information in a understandable form or an objective manner. Uncertainties were never discussed. Scare tactics warned parents about the dangers of asbestos in schools The idea that asbestos if not removed would kill school children was seemingly unassailable. EPA had never even prepared a creditable risk assessment on which to base its activity. And what became apparent in 1990, although it really was present earlier, was bitter controversy among scientists, and a great deal of legal action, public confusion, and exploitation. When EPA released The Five Facts and its fourth guidance document it was too little, too late. By 1990, vested interests in the legal profession (plaintiffs' lawyers), and rip out companies which often included those who performed inspections, had become entrenched. And they fought back with acrimonious and vituperous accusations claiming scientists who sought a change in policy by utilizing science and risk assessment were merely doing so because they were bought by the asbestos product manufacturers.

After billions of dollars had already been spent the policy articulated by AHERA and EPA was finally being questioned by a number of scientists. The economic cost was great. The cost to the scientific establishment was also great. Eminent authorities argued on both sides of the question. What had been referred to as the Selikoff Camp, scientists at the Mt. Sinai School of Medicine who had pressed for the removal of asbestos and believed that one fiber will kill you, were either unwilling or unable to reevaluate their position. Philip Landrigan and William Nicholson led the charge along with scientists in the Collegium Ramazzi against those scientists who reevaluated the potential harm from asbestos in school buildings and determined it to be not as dangerous as previously believed, and labeled EPA policy a fiasco. It is not easy to challenge conventional wisdom. Indignant and recriminating letters appeared in the scientific and popular press. The controversy eroded the image of impartial, objective scientists seeking the truth. It was politics as usual.

REFERENCES

1. 40 CFR, Part 763, 52 *Federal Register*, October 30, 1987, Sec. 2643 (dYi).
2. *Asbestos Hazard Emergency Response Act*, Sec. 201 (b)1.
3. *Asbestos Hazard Emergency Response Act*, Sec. 203(c)(d).
4. Findley, M.E., Rose, V.E., Cutter, G.E., and Windsor, R.A., Assessment of the EPA asbestos hazard algorithm, *Am. J. Publ. Health,* 73, 1179, 1983.
5. Perkins, J.L., *Modern Industrial Hygiene: Recognition and Evaluation of Chemical Agents*, Vol. I, Van Nostrand Reinhold, New York, 1997, chap. XIV.
6. *Federal Register*, Vol. 5, No. 83, Proposed rules, 15831, April 30, 1987.
7. USEPA, Office of Environmental Assessment, *Airborne Asbestos Health Assessment Update*, Washington, D.C., Dec. 1985. (EPA 600/8-84-003F).
8. Hughes, J.M. and Weill, H., Asbestos exposure-quantitative assessment of risk, *Am. Rev. Respiratory Dis.*, 133, 5, 1986.
9. Hearings Before Senate Subcommittee on Toxic Substances and Environmental Oversight, Committee on Environment and Public Works, 99 Congress, 2nd Session, May 15, 1986. (Testimony of John A. Moore, Assistant Administrator for Pesticides and Toxic Substances, EPA).
10. *Federal Register*, Vol. 52, No. 83, Proposed rules, 15838, April 30, 1987.
11. *Federal Register*, Vol. 52, No. 83, Proposed rules, 15838, April 30, 1987, 5875.
12. *Federal Register*, Vol. 52, No. 83, Proposed rules, 15838, April 30, 1987, 5875.
13. USEPA, *Study of Asbestos-Containing Materials in Public Buildings: A Report to Congress,* Washington, D.C., February 1988.
14. Heidhorn, R., Jr., Asbestos who's cleaning up, *The Philadelphia Inquirer*, Oct. 22–25, 1995.

15. Bernarde, M.A., *Asbestos the Hazardous Fiber*, CRC Press, Boca Raton, FL, 1990, 231.
16. *Federal Register*, Vol. 52, No. 83, Proposed rules, 15834, April 30, 1987.
17. *Wall Street Journal*, A16, Sept. 18, 1989. A16.
18. *Asbestos Monitor*, Vol. 2, #5, May, 1990, 1.
19. *Asbestos Monitor*, Vol. 2, #5, May, 1990, 1.
20. W.R. Grace Company and U.S. Gypsum litigation departments, unpublished data.
21. Comprehensive review of recent asbestos research, *N. Engl. J. Med.*, June 1989.
22. USEPA, *Asbestos-Containing Materials in School Buildings: A Guidance Document,* O.T.S., Washington, D.C., March, 1979.
23. Bernardi, M.A., *Asbestos the Hazardous Fiber*, CRC Press, Boca Raton, FL, 1990, 242.
24. Dillon, S., Asbestos task force members say confusion ruled testing, *New York Times*, B1, Sept. 13, 1993.
25. Dillon, S., Asbestos task force members say confusion ruled testing, *New York Times*, Sept. 13, 1993.
26. Dillon, S., A history of mistakes is cited for schools unit on asbestos, *New York Times*, A1, Aug. 12, 1993.
27. Dillon, S., Asbestos task force members say confusion ruled testing, *New York Times*, Sept. 13, 1993.
28. Dillon, S., School chancellor orders emergency asbestos lists, *New York Times*, B3, Sept. 14, 1993.
29. USEPA, *Managing Asbestos In Place: A Building Owner's Guide to Operations and Maintenance Programs for Asbestos-Containing Materials*, OPTS, July 1990.
30. Mossman, B.T., Bignon, J., Corn, M., Seaton, A., and Gee, J.B.L., Asbestos: scientific developments and implications for public policy, *Science,* 247, 294, 1990.
31. Mossman, B.T., Bignon, J., Corn, M., Seaton, A., and Gee, J.B.L., Asbestos: scientific developments and implications for public policy, *Science,* 247, 299, 1990.
32. Abelson, P., *Science,* 247, 1017, 1990.
33. Abelson, P., *Science,* 247, 1017, 1990.
34. USEPA, *Managing Asbestos In Place: A Building Owner's Guide to Operations and Maintenance Programs for Asbestos-Containing Materials*, OPTS, July 1990.
35. USEPA, *Managing Asbestos In Place: A Building Owner's Guide to Operations and Maintenance Programs for Asbestos-Containing Materials*, OPTS, July 1990, vii.

7 Dueling Conferences

INTRODUCTION

This chapter deals with the division within the scientific community over the hazards of asbestos and with the efforts of a part of that community, referred to as revisionists, to reevaluate the science and policy for asbestos in buildings. The reevaluation began with the questioning of the risk estimates, science and methodology of scientists at Mount Sinai, which the Environmental Protection Agency (EPA) utilized to develop its policy for asbestos in school buildings. The risk estimates of William Nicholson, a member of the Mount Sinai faculty, and the "one fiber can kill" theory had formed the basis for EPA guidance documents and asbestos-in-schools legislation such as The Asbestos Hazard Emergency Response Act (AHERA). It took almost a decade before the revisionists were heard. By 1990, knowledge of the excesses, economic waste, mismanagement, and misunderstanding associated with asbestos-in-schools policy finally reached the public.

After 1990, the federal government, i.e., EPA stated it favored management of asbestos-containing materials (ACM) in place, rather than removal, which was the previously prescribed method of dealing with asbestos in schools.[1] The effort to base policy for asbestos in schools on science and rational decision making began with questioning of both the risk estimates that EPA had relied on and the refusal of the agency to include measurements of asbestos-in-air of school buildings to determine how much asbestos occupants breathed. Measurement of asbestos-in-air could determine the level of risk, if any, to building occupants. EPA overlooked the fact that asbestos must be inhaled at concentrations in air deemed to be hazardous to health to cause illness. Today it is an accepted fact that the risks are minimal, and that removal of ACM is usually the least desirable option for dealing with asbestos in school buildings.[2]

THE AMERICAN INDUSTRIAL HYGIENE ASSOCIATION POLICY STATEMENT

When one of the most outspoken critics of federal asbestos-in-schools policy, and perhaps the best-known, addressed an audience of thousands of industrial hygienists at their annual meeting in 1985, he traced his own personal 30-year involvement with asbestos in the workplace and in the non-occupational environment. He sought some lessons for industrial hygienists who should have, but did not speak out about the manner in which asbestos policy evolved in the 1980s. He chided hygienists for refusing to speak out when EPA departed from the industrial hygiene paradigm of recognition, evaluation, and control of potentially hazardous material. Summing up he said,

> The movement of asbestos from the occupational to the non-occupational environment is a case study that will undoubtedly be followed in the future by other potentially toxic materials. We must determine our position as a profession with regard to the tools we utilize, the procedures we follow, and our understanding of our moral and legal obligations to prepare for often scientifically and technically insupportable positions assumed by others, including government agencies in their zeal to bring about change. We must discourage the opportunistic distortion of professional practices in these matters, and must issue well considered statements as a profession, statements relevant to the way these matters are addressed by society. We have a responsibility to lend perspective to these issues, to not permit understandably emotional responses to documented past severe health effects in other areas, such as the case of asbestos in the work place, to carry over into conditions of very low exposures in the public domain. We must remind people of the relevance of dose-response and toxicological principles to assessment of risk.[3]

It took the American Industrial Hygiene Association six years after this address to present a policy statement on asbestos in buildings consistent with the industrial hygiene paradigm. A summary of their policy statement of 1990 follows.

> The health risks associated with asbestos exposure for building occupants has been demonstrated to be very low. The decision to remove asbestos-containing materials (ACM) in undamaged, intact condition that are not readily accessible to occupants should be made only after assessing all other options. Both technical and financial issues should be fully explored by a team of trained specialists, including industrial hygienists, architects, and engineers. The optimal solution will vary from building to building, based on factors unique to each situation. One important consideration is the use of a well-designed air monitoring program to identify changes in airborne levels of asbestos. Special training and maintenance programs are needed to ensure the safety and health of building and contract workers who may encounter asbestos or who may disturb it during routine or non-routine activities.[4]

EVALUATION BY TWO MEMBERS OF THE SCIENTIFIC COMMUNITY

Others in the scientific community developed risk assessments for asbestos in buildings. But neither the popular press nor the public took much notice until 1990. The new risk assessments led to disease estimates for occupants of buildings with ACM that were uniformly minuscule, and contradicted those utilized by EPA. Most appeared in reports of commissions or scholarly journals, apparently seldom read by the public or Congressional aides. For example, in an important paper published in the *Annual Review of Public Health*, in 1986 (the same year Congress passed AHERA), Hans Weill and Janet Hughes raised questions about the extent of health risk to children from asbestos in school buildings.[5] The authors wrote:

> Optimal societal benefit from public policy decision making concerning environmental and occupational health issues can be obtained only when these decisions are fully and appropriately based on existing scientific knowledge. Regarding asbestos and its use in our society, this important goal has been illusive for a number of reasons. First, scientists have been unwilling or unable to communicate technical issues to decision makers, in part due to their complexity, but also because the forum for this interaction has been adversarial in nature. Second, the public and their policy makers have sometimes mistrusted science and technology; particularly recently, technological progress is often viewed as detrimental to society, particularly in environmental matters. And third, scientists opinions have, at times been based more on personal choice than on professional expertise.[6]

A fourth reason, but one not mentioned by Weill and Hughes, is that policy makers often choose to listen to scientists who supported the decision makers own point of view or who are, at that point in time, politically correct.

Discussing risk assessment in the same article, Weill and Hughes pointed out the ways researchers calculated risks in both the workplace and in schools and referred to the conclusions of the Ontario Royal Commission which had reviewed all available data and concluded that the best available estimate of the average asbestos concentrations inside buildings containing asbestos is 0.001 f/ml of air, with peaks of 0.01 f/ml occurring occasionally.[7] "One cannot escape the conclusions that the exposure of general building occupants to asbestos fiber concentrations in the air is negligible under most conditions; the conditions under which those exposures are not negligible are readily identifiable."[8] Weill and Hughes prepared a table of estimated lifetime risks from asbestos exposure in schools per one million students exposed at 0.001 f/ml of mixed fibers for five school years beginning at age ten, based on reports of several government agencies (EPA, CPSC, NRC, RCA, HSE), and their own reports concerned with potential risk from asbestos exposure, to indicate that scientific input should be part of the decision making process.[9] They noted also that derivation of potential risk estimates did not always proceed optimally. They illustrated their point utilizing the evaluation of EPA's risk estimates for school asbestos exposure. EPA risk assessment went through a lengthy process which included a series of drafts and many changes. The agency reduced its original risk estimate of 100 to 8,000 premature deaths attributable to asbestos in schools to a total of 40 to 400 in a later draft. In the final draft, EPA did not even present risk estimates. The agency had based all their estimates on the same exposure data.[10] The authors suggested that EPA resolve the details of risk estimation, such as model selection, parameter estimation, critical review of epidemiological data, and details of risk calculations before they issue a draft report. This was suggested to avoid the likelihood of using preliminary risk estimates as a basis for decision making after they have been superseded by subsequent versions.

EPA's series of draft reports with initially high estimates of potential risk created public uncertainty, fear, and in some cases panic. Many parents who had concluded that ACM in their children's schools presented a high cancer risk referred to the risk based on these early assumptions. Table 1 is based on a table Weill and Hughes developed. It indicates published estimates of risk in the United States from a variety

TABLE 1
Published Estimates of Risk From Various Causes[12]

	Annual rate (deaths per million)
Voluntary Risks	
Long-term smoking	1200
Bicycling ages 10–14 (1978)	14
High school football (1970–1980)	10
Eating 4 tbsp peanut butter/day (aflatoxin)	8
Aircraft Accidents (1979)	6
Living in a brick building rather than wood (radiation)	5
Whooping cough vaccination (1970–1980)	1–6
Asbestos in Schools (prevalent conditions)	.02–.37
Involuntary Risks	
Home Accidents (ages 1–14)	60
Motor Vehicle Accidents Pedestrian (ages 5–14)	32
Drowning (ages 5–14)	27
Fire (ages 5–14)	16
Inhalation/ingestion of foreign objects	15
Falls (ages 5–14)	4
Tornadoes (Midwest)	2
Floods	2

of causes. These estimates have been referred to in many documents because they indicate that far greater risks than asbestos in schools, from a variety of causes, both voluntary and involuntary, affect school children.

Weil and Hughes asked the following question, "... assuming an appropriate analysis results in risk estimates that are widely believed to compare favorably with the great number of voluntary and involuntary risks that we all face daily, is it a reasonable use of the resources of financially limited school boards to engage in wholesale replacement of asbestos-containing ceilings at a national cost of many billions of dollars?"[11] This questioning occurred in response to EPA's inability to utilize risk assessment for its asbestos-in-schools policy.

HEALTH EFFECTS INSTITUTE — ASBESTOS RESEARCH

In 1988, in an appropriations bill for EPA, Congress allotted two million dollars for asbestos research. The conference committee report stipulated that the research should be conducted under the auspices of the Health Effects Institute and formed the Health Effects Institute-Asbestos Research (HEI-AR) in response to a mandate from Congress of August 3, 1988. HEI-AR was asked to perform the following tasks: (1) to determine the actual airborne asbestos levels prevalent in buildings; (2) to characterize peak exposure episodes and their significance for potential adverse health effects on exposed building occupants; and (3) to evaluate the effectiveness of asbestos management and abatement strategies in a scientifically meaningful manner.[13] Congress established HEI-AR as an independent, nonprofit organization. It received financial support from a broad range of groups with interest in asbestos. An allocation of funds to support the effort specified private sector funding of two million dollars each year for at least three years as follows: (1) $633,333 per year from present and former asbestos manufacturers; (2) $633,333 per year from real estate industry sources, i.e., realtors, developers, building owners and managers, mortgage bankers; (3) $633,333 per year from insurance industry sources; and (4) $100,000 per year from all other interested organizations including labor unions and environmental groups; $2,000,000 total to be matched on a 50-50 basis with federal funds.[14]

Beginning in 1989, the HEI-AR made several contributions to understanding the risks associated with indoor asbestos. In 1991, HEI-AR published a report entitled, *Asbestos in Public and Commercial Buildings*.[15] It was a synthesis of evidence on asbestos in buildings and associated disease rates. Table 2 lists the Board of Directors of the HEI-AR, members of the Research Oversight Committee, members of the HEI-AR Asbestos Review Panel, and their affiliations. Recurring issues in the 1991 HEI-AR report included: fiber type, asbestos type, the question of whether or not a threshold existed for cancer and/or mesothelioma, and questions about the level of building occupant exposure to asbestos in buildings. All of the above issues were not new. Nevertheless they required clarification. The report, in order to do so, reviewed all literature to determine what was known.

The challenge to the Literature Review Panel, a multidisciplinary group of experts, was tc gather all research on asbestos and to generate reliable, objective information. Specifically, to, "determine actual airborne (asbestos fiber) levels prevalent in buildings; characterize peak exposure episodes and their significance, and to evaluate strategies in a scientifically meaningful manner."[17] The purpose of the report was to, "review and synthesize the state of knowledge as reflected in scientific articles, reports and additional unpublished data on four issues considered pertinent to the congressional mandate."[18] The four issues were:

1. The concentrations of airborne asbestos fibers found in public and commercial buildings;
2. The concentrations of such fibers to which building occupants, including custodial workers, maintenance workers, abatement workers, and other occupants are exposed; the situations causing such exposures; and the potential for adverse health effects resulting therefrom;
3. The possible impact that different asbestos remediation strategies may have on the exposure of building occupants to airborne asbestos and, in turn, on the risks of health effects in those exposed;
4. The significance of each from of asbestos in terms of its potential ill health effects and its implications for different remediation options in buildings.[19]

The HEI-AR report addressed specific topics such as the assessment of asbestos exposure and remediation of ACM. But the largest section reviewed existing scientific literature on health implications.

TABLE 2[16]

Board of Directors	
Archibald Cox, Chairman of the Board	Harvard Law School, Emeritus
William O. Baker	Bell Laboratories, Emeritus
Donald Kennedy	President, Stanford University
Charles W. Powers	Resources for Responsible Management
Research Oversight Committee	
Jonathan Samet	University of New Mexico
Carl Barrett	National Institute of Environ. Health Sciences
John M.G. Davis	Institute for Occupational Medicine, Edinburgh
David G. Hoel	National Institute of Environmental Health Sciences
Morton Lippmann	New York University
Julian Peto	Institute of Cancer Research, London
Literature Review Panel	
Arthur C. Upton, Chairman	New York University
J. Carl Barrett, ex officio member	National Institute of Environ. Health Sciences
Margaret R. Becklake	McGill University, Montreal
Garry Burdett	U.K. Health and Safety Executive
Eric Chatfield	Chatfield Technical Consultants
John M.G. Davis	Institute for Occupational Medicine, Edinburgh
Gordon Gamsu	University of California
David G. Hoel, ex officio member	National Institute of Environmental Health Sciences
Arthur Langer	Brooklyn College
Richard J. Lee	R.J. Lee Group
Morton Lippmann	New York University
Brooke T. Mossman	University of Vermont
Roger Morse	ENTEK Environmental Services
William Nicholson	Mt. Sinai School of Medicine
Julian Peto	Institute of Cancer Research, London
Jonathan Samet, ex officio member	University of New Mexico
J. Chris Wagner (retired)	Dorset
Patrick Kinney, Consultant	New York University

With regard to the question of assessment of asbestos exposure, the panel reviewed what was known about the amount of asbestos in air people breathed. They reviewed various analytical methods to measure asbestos in air. (The amount, methods by which the sample is collected and measured on air collected on filter samples can have a profound effect on the final result, expressed as asbestos fibers per unit volume of air, either total fibers or fibers greater than a certain length.) The main issue in analysis was whether to use a direct or an indirect method. The indirect method sonicates the sample and breaks larger fibers and matrices of fibers into smaller fibers, thus increasing the reported numerical value of the air concentration. A related issue was whether all fibers, regardless of length, have equal carcinogenic potential. Fibers larger than approximately five microns are believed to be biologically active; fibers smaller than five microns are easily cleared from the human lung by various biological mechanisms. Another issue related to asbestos exposure reviewed by the panel was, what are the concentrations of asbestos-in-air to which occupants of buildings and schools are exposed?

Also of concern was exposure to asbestos of maintenance personnel, individuals who work in close proximity to asbestos containing products above the ceilings and in the boiler rooms of schools and commercial buildings. The scientists who wrote the report prepared tables summarizing relevant data from peer reviewed scientific literature, government reports, and the results of unpublished studies utilized in asbestos litigation.

The members of the asbestos literature review panel investigated remediation of ACM in buildings which was also a cause of contention. They asked, "Should asbestos-containing materials in buildings be left in place and be carefully managed to minimize any exposure?" Or, "Should it be removed?" They summarized the known information about the efficacy of administration and the various work practices to minimize fiber release and subsequent exposure. They discussed the merits of encapsulation, which involves spraying a liquid material on the ACM. The sprayed liquid is absorbed by the ACM causing cohesion of contents and preventing fiber release from bulk asbestos-containing material. The panel found documented problems with encapsulation as well as with removal. In some cases the weight of the encapsulant caused the ACM to break off into sections and fall to the floor. Work practices required supervision, worker instruction, and adherence to prescribed procedures at all times.

Removal involved breaking the asbestos-containing bulk matrix, causing the release of large quantities of airborne fibers which had to be contained. If they were not contained the fibers would circulate throughout the rest of the building. Within the containment area, during removal, high fiber concentrations demanded stringent use of personal protective equipment and work practices by workers. Because this seldom occurred, workers were often exposed to significant amounts of asbestos dust. Removal costs could average from 10 to 20 dollars per square foot of ACM.

It is easy to imagine the difficulty of making decisions in each particular situation. For example, if the ACM was in poor condition, with future potential for fiber release, removal might be the best option. But, if the material was in good to excellent condition, with little access by personnel, and with low potential for fiber release, managing the ACM in place would be the best option. (Managing in place is also referred to as O and M, or Operations and Maintenance.)

The largest section of the literature review examined existing scientific literature on the known manifestations of human asbestos disease,* including asbestosis, pleural diseases, lung cancer, pleural and peritoneal mesothelioma, other cancers, and benign lung masses. They placed emphasis on exposure/risk relationships, where available.

In the final chapter, the panel estimated the lifetime cancer risks for airborne exposure to asbestos fibers. They concluded asbestosis would not be anticipated

* Chapter 2 deals with much of this information.

TABLE 3[20]
Estimated Lifetime Cancer Risks for Different Scenarios of Exposure to Airborne Asbestos Fibers[a]

Conditions	Premature cancer deaths (lifetime risks) per million exposures persons
Lifetime, continuous outdoor exposure	
• 0.00001 f/ml from birth (rural)	4
• 0.0001 f/ml from birth (high urban)	40
Exposure in a school containing ACM, from age 5 to 18 years (180 days/year, 5 hours/day)	
• 0.0005 f/ml (average)[b]	6
• 0.005 f/ml (high)[b]	60
Exposure in a public building containing ACM age 25 to 45 years (240 days/year, 8 hours/day)	
• 0.0002 f/ml (average)[b]	4
• 0.002 f/ml (high)[b]	40
Occupational exposure from age 25 to 45	
• 0.1 f/ml (current occupational levels)[c]	2,000
• 10 f/ml (historical industrial exposures)	2,000,000

[a] This table represents the combined risk (average for males and females) estimated for lung cancer and mesothelioma for building occupants exposed to airborne asbestos fibers under the circumstances specified. These estimates should be interpreted with caution because of the reservations concerning the reliability of the estimates of average levels and of the risk assessment models summarized in Chapter 8.
[b] The "average" levels for the sampled schools and buildings represent the means of building averages for the buildings reviewed herein (Figure 8.2; Table 8-2). The high levels for schools and public buildings, shown as 10 times the average, are approximately equal to the average airborne levels of asbestos recorded in approximately 5% of schools and buildings with asbestos-containing materials (ACM) (see Chapters 4 and 8). If the single highest sample value were excluded from calculation of the average indoor asbestos concentration in public and commercial buildings, the average value is reduced from 0.00020 to 0.00008 f/ml, and the lifetime risk is approximately halved.
[c] The concentration shown (0.1 f/ml) represents the permissible exposure limit (PEL) proposed by the U.S. Occupational Safety and Health Administration. Actual worker exposure, expected to be lower, will depend on a variety of factors including work practices, and use and efficiency of respiratory protective equipment.

from the low level exposures in buildings. They derived estimates for cancer risks using the linear no threshold extrapolation model and the exposure concentrations and durations of exposure for persons in buildings which emerged from their review. A variety of scenarios was hypothesized (Table 3).

The writers of the HEI-AR report summarized their conclusions. Regarding control of asbestos exposure they noted that although limited, the data supported the following generalizations. ACM in buildings in good repair and undisturbed is unlikely to give rise to airborne asbestos fiber concentrations above levels found outside those buildings. Accessible ACM has potential to be damaged. During processes that damage the ACM fibers can be released into the air. Maintenance activities can result in localized increases in airborne asbestos levels exposing the workers involved and possibly nearby building occupants. Operations and maintenance procedures can reduce those exposures. Removal of ACM from buildings, if improperly done, can cause increases in airborne fiber levels.[21] The executive summary reinforced the idea that if a building is well-maintained with long-term airborne levels of asbestos fibers similar to ambient background levels, removal or other abatement action, if improperly done, can cause building fiber concentrations to increase.[22]

The problems associated with operations and maintenance programs proved so complicated that HEI-AR organized a separate workshop on this subject on March 8–9, 1993 in Cambridge, Massachusetts. At the conference participants discussed asbestos management programs and considered both the usefulness and limitations of the operations and maintenance approach.[23]

THE HARVARD SYMPOSIUM

Before publication of the HEI-AR literature review, the Harvard University, Energy and Environmental Policy Center, John F. Kennedy School of Government hosted a "Symposium on Health Aspects of Exposure to Asbestos in Buildings." The symposium participants met on December 14–16, 1988. Prior to this meeting, there had been neither a conference, symposium, or written compendium summarizing the current state of knowledge of the health risks of asbestos, if any, associated with asbestos in buildings. The Harvard symposium was particularly timely. It met at the same time Congress deliberated the extension of AHERA to all commercial and public buildings. A great deal of scientific and technical data and insight had been gained during the decade of the 1980s, but there had not yet been a forum to discuss the implications of these findings for public policy, as contrasted to their relationship to the advancement of scientific understanding which had already occurred through the mechanism of publication in scientific and technical journals. Until the Harvard Symposium, recent research on asbestos health effects had not been integrated into public policy options. The Harvard Symposium proposed to do this.

As one might expect, sponsors of the Symposium included groups anticipating a major impact on their members, if Congress passed a law for commercial buildings similar to The Asbestos Hazard Emergency Response Act of 1986. The Safe Buildings Alliance, The National Association of Realtors, The Institute of Real Estate Management Foundation, and The Urban Land Institute sponsored the symposium. The Symposium organizers maintained that sponsors allowed The Energy and Environmental Policy Center of Harvard University's John F. Kennedy School of Government a free hand in structuring the content of the symposium and its proceedings.[24] The participants and their affiliations are indicated in Table 4 entitled, "Participant List and Affiliation, Harvard Symposium on Health Aspects of Exposure to Asbestos in Buildings." The participants included people with diverse viewpoints on the issue of asbestos in buildings.

In order to address the efficacy of removing asbestos from buildings participants first needed to reassess whether the extent and magnitude of the hazard was effectively measured and if the risks were large enough to warrant concern. Those potentially at risk included both the occupants and workers who remove asbestos. Summarizers of the meeting wrote the following

> Ultimately, we must be confident that removing asbestos fibers either now or prior to demolition will result in a reduction of health risk. In light of scientific evidence published since the National Research Councils' 1984 report on asbestos, the experience with current removal practices and low level of demonstrated asbestos fiber exposures in buildings, and new information on potential hazards of man-made mineral fibers used as substitute insulation, there is a reasonable conclusion that removal of asbestos may actually increase health risk. Given this new information it is appropriate to reappraise the credibility of the scientific database and assertions made with respect to the health risk of asbestos in buildings."[26]

The symposium addressed a number of questions. For example: What is the extent of the health risk posed by asbestos in buildings? Who in the population is being exposed? How do we measure the level of exposure? How do we communicate the extent of risk? Participants discussed methods for assessing concentrations and exposures. The executive summary listed and summarized principal findings. It is paraphrased as follows:

1. Asbestos is not a single fiber. Differences in fiber type and dimension can be of significance in determining health risks. For example, amphibole fibers pose a greater risk of mesothelioma than chrysotile fibers. Respirable fibers longer than approximately 5 μm are thought to be of much greater risk than short fibers.
2. Occupants' risk should be determined by exposures to airborne fibers. The consensus view holds that air sampling, using the direct preparation TEM methodology, is the method of choice for current exposure assessment. The available asbestos air monitoring database for United States buildings indicates extremely low average concentrations of airborne asbestos under normal building use conditions.
3. The extent of risk associated with exposure to asbestos differs for various groups who work or reside in the building. Custodial workers, maintenance and construction workers may be exposed to elevated levels of asbestos for brief periods of time. Most office workers, teachers, students and other building occupants typically do not come in close contact with asbestos-containing materials.
4. Considerable differences exist in the value of many early studies on health effects of asbestos. They contain differences in measurement techniques and many uncertainties.
5. Process factors and environmental effects have potentially significant impact upon risk. Relative risk of lung cancer from exposures to amosite or mixed (amphibole and chrysotile) fibers is typically higher than risk associated with processes in which chrysotile is bonded with other materials.
6. The majority of asbestos-related mesotheliomas can be attributed to amphiboles (primarily crocidolite), much less commonly used in the

TABLE 4
Participant List and Affiliation[25]

Harvard Symposium on Health Aspects of Exposure to Asbestos in Buildings

Charles Achilles
Institute of Real Estate Management Foundation

Tom Black
Urban Land Institute

Jean Chessson
Chesson Consulting

Mort Corn
Johns Hopkins University

Donald Dewees
University of Toronto, Canada

Sandra Eberle
Consumer Product Safety Commission

Nurtan Esmen
University of Pittsburgh

J. Bernard Gee
Yale University School of Medicine

Michael Gough
Center for Risk Management

Timothy Hardy
Kirkland and Ellis

Dick Hopper
U.S.G. Corporation

Sarah Hospedor
National Institute of Realtors

Janet M. Hughes
Tulane University

Richard Innes
Legislative Assistant to Senator John Chaffee

Patrick Kinney
Health Effects Institute

Arthur Langer
Brooklyn College of the City University of New York

Si Duk Lee
Harvard University
U.S. Environmental Protection Agency

Peter Lees
Johns Hopkins University

Sue Lin Lewis
Harvard University

Douglas Liddell
McGill University, Canada

John F. McCarthy
Environmental Health and Engineering, Inc.

Karen Millne
U.S. Environmental Protection Agency

Kenneth Millian
W.R. Grace and Co.

Brooke Mossman
University of Vermont

Halûk Ozkaynak
Harvard University

Edward Peters
Arthur D. Little, Inc.

Julian Peto
Institute of Cancer Research, UK

Charles Powers
Health Effects Institute

Bertram Price
Price Associates

Robert Sawyer
Entke, Inc.

James Smith
Georgia Institute of Technology

John D. Spengler
Harvard School of Public Health

Edlu Thom
W.R. Grace and Co.

Hans Weill
Tulane University

Dietrich Weyel
University of Pittsburgh

John Welch
Safe Buildings Alliance

Richard Wilson
Harvard University

Mark Wine
Safe Buildings Alliance

United States. The mesothelioma risk from exposure to chrysotile asbestos, the type of asbestos most commonly found in United States schools and buildings, is believed to be considerably lower.

7. Lung cancer and mesothelioma risk models have been developed for asbestos exposure, which are considered conservative because they tend to overpredict rather than underpredict the risks. Recent data indicates the average concentration of asbestos in schools and other buildings with asbestos-containing materials is generally below 0.001 (mixed) f/ml used in risk calculations for school children. Thus, using conservative risk models and exposures higher than typically measured, the projected lifetime risk from exposure to mixed asbestos fibers is one death among a cohort of 100,000 children.
8. The risk of 1 in 100,000 is far less than most commonly experienced environmental health risks, for example, tobacco smoke and radon. Although risks posed by in-place asbestos is small in absolute and relative terms, public perception of these risks is high, often leading to the decision to simply remove all asbestos-containing materials.
9. Removal has risks. Removal and disposal exposes the removers to high concentrations of airborne asbestos. It can potentially increase, rather than decrease indoor concentrations of asbestos-in-air.
10. Spending money on asbestos removal, in many cases, will likely decrease available funding to support other public health and educational measures that could be more effective in reducing other environmental health risks.
11. Air sampling and risk calculations should be performed as a component of selection of an effective management strategy for health protection and risk reduction. Evaluation should include risks to occupants of buildings, custodial and maintenance personnel and asbestos removal workers. Potential for exposure to asbestos fibers is much greater for the latter group than for occupants.[27]

The conclusions of the Harvard Symposium contradicted the Federal government and EPA policy reflected in the AHERA rulemaking. The Symposium concluded that expensive remedial measures taken to reduce exposure to asbestos in American schools was unwarranted, unnecessary, and downright dangerous. The HEI-AR report and the Harvard Symposium drew harsh criticism from labor unions, groups acting on behalf of asbestos victims, plaintiff's lawyers, the asbestos abatement community, EPA, and some members of the scientific community.

Old policies do not go quietly into the night. Especially when the groups who accept them have interests dependent on those policies. When the HEI-AR Report became public the *Wall Street Journal* quoted two scientists on the faculty of Mount Sinai Medical School. Philip Landrigan, a pediatrician and chair of the New York State Task Force on Asbestos in Buildings, called the HEI-AR Report a white wash. William Nicholson, also on the faculty of Mount Sinai School of Medicine and a member of the committee which wrote the HEI-AR report, said he planned to present a dissenting statement. Nicholson, quoted in the *Wall Street Journal* said, "The report can be interpreted to imply there is virtually no asbestos in buildings in the United States, and no reason to be concerned about it."[28]

If the HEI-AR Report and the Harvard Symposium conclusions were correct, the billion dollar well-organized asbestos industry and the plaintiffs and their lawyers who sued manufacturers, installers, and insurers of ACM for huge sums of money would lose a great deal. Asbestos removal and legal suits to recover costs of removal had become a money making bonanza. The issue of public health faded fast when confronted with making large sums of money. As noted earlier in this chapter, both the report and Symposium appeared at the same time Congress considered extending AHERA to commercial buildings. EPA had already estimated extension of their requirements under AHERA to approximately 733,000 public and commercial buildings containing asbestos would cost 53 billion dollars, a figure considered an underestimate of the cost. The reports led to fierce and vituperative controversy fed by both sides and reflected in the growing number of articles about the controversy, which had attracted media attention.

SCIENTISTS QUESTION POLICY

An increasing number of scientists voiced their concerns about the extraordinarily wasteful use of public resources they believed had been squandered on a non-problem, based on fear, not scientific data. Activity related to asbestos now was labeled the "asbestos panic." Many believed that unnecessary removals had turned into a money making frenzy. Finally, the scientific community was openly questioning the wisdom of EPA's policy for ACM in schools, and the resulting expenditure of public funds for a scientifically insupportable concept, i.e., extremely low levels of exposure to asbestos in school buildings can cause a major public health hazard.

By 1990, other noteworthy documents countered the notion that ACM in schools would cause a new epidemic of lung cancer and mesothelioma.[29] The most influential was published in the January 19, 1990 edition of *Science*. The article entitled, "Asbestos: Scientific Developments and Implications for Public Policy"[30] was written by five scientists, three from the United States, one from France, and one from Scotland. The authors summarized recent developments in asbestos research and discussed their implications for public policy. They posed the following questions: Does available evidence support the concept that asbestos causes disease in the non-occupational environment? What are the mechanisms of asbestos-induced fibrogenesis and carcinogenesis? Most importantly, have recent data been adequately considered in formulating policies in the United States for regulation and banning of asbestos?[31] They concluded that available data do not support the theory that low-level exposure to asbestos poses a health hazard in schools and buildings. "... The available data do not indicate that asbestos associated malignancies or functional impairment will occur as a result of exposure to most airborne concentrations of asbestos in buildings."[32] They said the type of asbestos found in most buildings was chrysotile asbestos, not the more potent amphibole types, and they reviewed recent studies which indicated that severity of asbestosis, incidence of carcinoma of the lung, and mesothelioma correlate with a lung burden of crocidolite and amosite asbestos, with amphibole types, but did not correlate with chrysotile asbestos. They cited studies that suggest amphiboles are more potent than chrysotile in induction of fibrotic lung disease and lung cancer.[33] They also indicated that epidemiological and experimental data supported the concept of a threshold for chrysotile induced pulmonary fibrosis.[34]

The *Science* article ended with a comment on the "Asbestos Panic," fueled said the authors by insupportable concepts such as the "one fiber theory" which maintained that one fiber of asbestos, if inhaled will cause cancer. They said the asbestos panic led to public pressure and haphazard removal of ACM from schools and public buildings. And the removal led to increased airborne concentrations of the fiber, sometimes for months after. Safe removal and disposal became a problem. Asbestos abatement also led, said the article, to exposure of a large, new unprotected group of relatively young asbestos removal workers, often exposed under suboptimal conditions. Under AHERA, which does not require or set standards for removal of asbestos, schools must submit a management plan detailing how they plan to deal with asbestos, often done with little expert advise. Schools could be fined up to $5,000 a day for lack of compliance or adherence to deadlines. The authors pointed out that EPA recommended bulk sampling of ACM to determine the presence of asbestos and a visual inspection to determine their course of action, rather than measurement of airborne levels of fibers. But, according to the authors, measurement of airborne levels of fiber are far more important to determine the need, if any, for removal of ACM.[35] In a conclusion to the article its authors wrote, "The available data and comparative risk assessments indicate that chrysotile asbestos, the type of fiber predominantly in United States schools and buildings is not a health risk in the non-occupational environment. Clearly, the asbestos panic in the United States must be curtailed, especially because unwarranted and poorly controlled asbestos abatement results in unnecessary risks to young removal workers who may develop asbestos-related concerns in later decades."[36]

Although much of what was written in the *Science* article had been said before, this publication received a large amount of attention in the media. It became a catalyst for change and caused a reevaluation of asbestos policy. Generally, the media does not respond to scientific findings except perhaps when a major breakthrough occurs, which might bring fame to a scientist or to the institution where it occurred. The press had been aware of the conclusions of the Harvard Symposium and published some editorials on the subject as early as September 1989.[37] Media attention climaxed at the beginning of 1990 after publication of the *Science* article. The full proceedings of the Harvard Symposium had also been published in 1990. Nevertheless, it was unusual for the press to keep reporting even weeks after the event. Articles appeared in *The New York Times*, "Risk is Seen In Needless Removal of Asbestos";[38] *The Wall Street Journal*, "Health Risks of Asbestos Downplayed";[39] *Time Magazine*, "An Overblown Asbestos Scare";[40] *Readers Digest*, "Great Asbestos Rip-Off";[41] and *Forbes Magazine*, "Paratoxocology."[42]

Why did such a response occur? Apparently the media was taken in earlier. Perhaps they felt that they had been fooled or used. Whatever the reason, the media reversed itself. Newspaper articles referred to a widespread unjustified panic that needed to be curtailed, or a hoax perpetuated on the American public. They used the phrase, too much money chasing too little risk to describe the fact that although health risk from ACM in schools and other buildings was minimal, money had been squandered for no good reason, especially since schools lacked adequate resources to solve other more pressing and real problems. *Forbes Magazine* published an article entitled, "Paratoxicology": "Scientists are convinced that asbestos in buildings isn't so dangerous after all. Why are we spending 5 billion dollars a year to rip it out? good question." The *Forbes* article summed up the public outcry, using the phrase public hysteria to describe reaction to asbestos in schools. *Forbes* claimed public hysteria about the asbestos threat reached its zenith when Congress passed AHERA, although some scientists who had been studying effects of low-level exposure to asbestos were convinced Congress had overreacted.[43] An EPA study confirmed the scientists' belief when it found asbestos fiber levels indoors no more dangerous than those outdoors.[44] "We live in a risky world, filled with serious threats, but low level exposure to asbestos in buildings does not seem to be one of them. Nevertheless fear of asbestos has created a $3 billion-a-year industry for lawyers, consultants, smart real estate developers — and especially for the so-called asbestos abatement contractors who get paid a lot of money to rip asbestos out of buildings and bury it in landfills."[45] Other points made in the article were: The basis for fear of exposure to asbestos is not founded in reality. No evidence exists that environmental exposure to asbestos is a public health hazard. The evidence used to create fear of asbestos, is "paratoxicology." The data is suspect since anticipated mortality rates turned out to be small compared to other everyday risks. The "one fiber can kill" theory is wrong. Strong evidence exists to prove some fibers are more deadly than others. Asbestos abatement usually raises levels of asbestos fibers in a building and endangers abatement workers. And asbestos is a manageable risk that need not be abated especially since removal can be more dangerous than leaving it alone.[46] This article and others reiterated findings found in the HEI-AR Report, the Harvard Symposium, and the *Science* article. The message here and in the popular press was a far different one from what had been presented during the 1980s leading up to AHERA. Why did the media do an about face and in the 1990s conclude that time, money, and energy were being squandered for a non-problem?

By the 1990s, the idea of relative risk had become part of the public understanding of environmental risk and in general, public perception of risk to children in schools changed. Asbestos was no longer perceived as a killer of school children. Yet, there remained people who could or would not reevaluate their position in light of the new scientific information. Cognizant that risks to school children from ACM in schools had been reassessed they responded with anger and recrimination, and a concerted effort to discredit the scientists whose research had challenged the old asbestos and disease paradigm. Rather than refute the science they impugned those scientists who questioned the exaggerated risks attributed to in-place asbestos and the removal of asbestos from school buildings. The conflict became downright vicious, as the controversy continued. For almost 20 years, the idea that asbestos, once a tragedy for workers, threatened school children and the general public was relatively unassailed. By 1990, challenges to the conventional wisdom about asbestos in schools and buildings came fast and from many directions. It captured the attention of the American public. The seemingly unassailable underlying concepts upon which asbestos policy rested, i.e., one fiber can kill, all asbestos types are equally hazardous, if its there remove it and if its a killer in the workplace then it must be equally dangerous in the non-occupational environment heretofore went unquestioned except in a few instances.

COUNTER ATTACK AGAINST THE REVISIONISTS

What can be termed as a counter attack began soon after the publication of the *Science* article. Members of the abatement industry voiced their concern in the trade journal, *Asbestos Abatement.* For example, an editorial written by a member who feared the new evidence discrediting the abatement industry said, "Few can deny that the public relations campaign of the Safe Buildings Alliance (SBA) has dealt the asbestos abatement industry another blow. The SBA has asserted that the abatement activities generate more harmful health effects than simply leaving asbestos-containing material alone. In addition, they take the position that asbestos is not dangerous in many situations. However, the medical and scientific evidence that exists disputes the SBA's claims."[47] In the same document, the president of the Asbestos Abatement Council, which published *Asbestos Abatement*, said, "you have probably heard of the so-called Harvard Report and the Safe Buildings Alliance (SBA). You've heard their propaganda and seen their effectiveness."[48] Clearly he meant to discredit the research presented at the Harvard Symposium as mere propaganda. The word "disinformation" appeared over and over as a response to the challenges of new information. The removal industry spokespeople did not want to accept the fact that asbestos in place did not pose as great a threat to people in buildings (low exposure) as asbestos did to workers (high exposure). Discrediting the scientists who sought change offered an outlet. Rather than deal with the message they attacked the messenger. The charge most often leveled at the scientists who criticized the regulations for failing to distinguish among threats posed by different types of asbestos, and who sought to change policy, was that the data the scientists presented was nothing more than the ideas of the asbestos industry, ideas which suggested that the danger from asbestos had been exaggerated. They implied and often said outright that the scientists had undisclosed conflicts of interest.[49]

DUELING CONFERENCES — THE THIRD WAVE

It did not take long before another conference was organized in order to rebut the evidence presented at the Harvard Symposium, the HEI-AR Report, and in the *Science* article. In response to those in the scientific community who argued that asbestos in buildings presents minimal hazards to health the Collegium Ramazzini, an international society of environmental and occupational health scientists, sponsored a conference entitled "The Third Wave of Asbestos Disease: Exposure to Asbestos In Place, Public Health Control." The organizers said they feared the debate over asbestos health risks threatened efforts to contain known asbestos hazards, and sought to heighten concern about the dangers of asbestos in buildings. In a six-page press release, four months before the conference, Irving Selikoff, the best known spokesperson associated with the "one fiber" theory and the fear of asbestos in buildings, presented the rationale for the conference. Selikoff said that a new phase of asbestos exposure had recently began, called the third phase. The first phase was associated with mining and the manufacture of asbestos products. It began in the second decade of the 20th century and continued until the 1980s. Some disease associated with these activities still occurs. However, Selikoff said it has decreased considerably with time. Phase two is marked by the use of asbestos products. Insulation workers and shipyard workers during World War II were exposed to asbestos. The third phase involves exposure to asbestos in place.[50] The third phase was explained in a different manner by Philip Landrigan in the proceedings of the conference, published in 1991. Landrigan defined phases one and two in the same manner as Selikoff, but he explained phase three differently. He wrote:

> Today, as buildings constructed with asbestos over the past six decades begin to age and deteriorate, serious potential exists for a third phase of asbestos disease among persons engaged in repair, renovation and demolition of these buildings. Potential also exists for serious environmental exposure to asbestos among residents, tenants and users of these buildings, such as school children, office workers, maintenance workers, and the general public. The Centers for Disease Control, the American Academy of Pediatrics, and the U.S. Environmental Protection Agency have projected that over the next 30 years approximately 1,000 cases of mesothelioma and lung cancer will occur among persons in the United States exposed to asbestos in school buildings as school children.[51]*

The "Third Wave" conference was held in New York City, June 6, 7, and 8, 1990. The New York Academy of Sciences published the proceedings in 1991.[54] The conference organizers received financial assistance from a plaintiff's fund granted through the Philadelphia Federal District Court and the Workplace Health fund. The New York Academy published a disclaimer at the beginning of the volume it published: "The New York Academy of Sciences believes it has a responsibility to provide an open forum for discussion of scientific questions. The positions taken by the participants in the reported conference are their own and not necessarily those of the Academy. The Academy has no intent to influence legislation by providing such forums."[53] The conference, on the other hand, meant to influence legislation, even though that idea went unstated. Environmental and occupational health scientists and researchers met alongside of lawyers, government officials, business and labor representatives, and judges to discuss "The Third Wave of Asbestos Disease." The day before the "Third Wave" meeting commenced in New York City, the U.S. Congressional Representative from Pennsylvania, Mr. Gaydos, spoke before the House of Representatives about the third wave of asbestos disease. His agenda was to extend AHERA to buildings other than schools.[54]

The conference was focused on a number of areas to demonstrate the severity of the problem of asbestos exposure. A significant part of the meeting highlighted mechanisms of disease, i.e., the way in which asbestos affects living organisms and in particular exposed human populations. A set of papers dwelt on mechanisms of

* These facts and figures were not documented by Landrigan.

action, in other words, how mutagenic and carcinogenic action progresses from the first mutagenic change to subsequent cellular alteration.[55] Presenters at the conference placed emphasis on dose response in an attempt to focus on effects of low level exposure by stressing the uncertainties of the dose response relationship in the low range.[56]

Curiously, although the conference was organized to respond to the concern for asbestos in public buildings and to low-level exposure, only 12 papers out of 57 specifically addressed asbestos in buildings. Three discussed children or teachers exposed to asbestos in place and twelve covered the topic of occupationally-exposed populations. Another area of emphasis in the proceedings, chrysotile and its carcinogenic properties, concentrated on the debate as to whether or not chrysotile fibers can cause mesothelioma, as contrasted to amphibole fibers. Radiological and physiological techniques for detecting abnormalities or papers to prove the usefulness of these techniques in epidemiological studies accounted for nine presentations.[57] Thus, a great deal of the conference did not concern the stated interest of the conference, but drew upon well known material (not new research) to justify grave fear for the health of occupants of buildings.

The most upsetting aspect of the "Third Wave" conference concerned the absence and response of scientists who worked in the area of non-occupational exposure to asbestos and had made contributions, but were nevertheless excluded from the meeting and although not present, were criticized and maligned. A number of letters written between November and December 1990 illustrate the point.

Bruce Case, Director of the Center for Environmental Epidemiology at the Graduate School of Public Health, University of Pittsburgh, wrote to Philip Landrigan, Conference Chair and Editor for the proceedings concerning the "Third Wave" conference and proceedings.[58] Case was concerned about the hostile and controversial atmosphere at the Third Wave meeting. He wrote the letter in reply to a request from Landrigan for amplification of material Case had presented at the conference, for publication on the proceedings. Case held different views from many of the people at the meeting and Landrigan had asked Case to submit a piece of whatever length Case chose in order to make his views known to readers of the Proceedings. In his letter, Case brought up other issues of contention. Angered because scientists with views different from the organizers' views were not welcome or invited to the conference, Case wrote, "I could name many scientists specializing in the area of asbestos-related research who were either excluded from the meeting or invited in a way that made it clear that they were not welcome. Others were invited; some (myself included) were present. Unfortunately, the atmosphere created prior to and at this meeting was so hostile that many scientists having other points of view, or simply a neutral informed approach, felt constrained to remain absent or silent."[59]

As an example of the kind of atmosphere that concerned him, Case quoted a newsletter entitled *Asbestos Watch.*

> The conference challenged the asbestos industry doctors, particularly Mossman, Gee and Corn to come forth and defend their pseudo-science white wash of the hazards of asbestos. Hundreds of scientists and others gathered to hear 56 presentations from world known physicians and scientists rebuking the industry propaganda about the

> safety of chrysotile, etc. Although invited, the asbestos-docs reported they could not come as they were too busy taking their blood money to the bank and feared the bright light of scientific investigation of their work ... the papers of the "Third Wave" conference will be published in the future, however, Dr. Selikoff allowed the White Lung Association of New Jersey to videotape the conference. Copies of the complete set of video tapes of this most important conference are available from the White Lung Association, P.O. Box 1483, Baltimore, MD, 21203-1483.[60]

Case suggested to Landrigan that he separate the scientific papers from the biased process of the meeting itself. It could be done he said, and offered a few suggestions. For example:

1. Peer review and comment. While certain the Academy will deal with the former, Case said the latter was equally important given the absence from the "Third Wave" conference of so many experts in the field. He suggested commentaries from absent experts such as Sir Richard Doll, Dr. Brook Mossman, Dr. Janet Hughes, Dr. Morton Corn, Dr. Bernard Gee, Dr. Jean Bignon, Dr. Hans Weil, Dr. Andrew Churg, Dr. Jerrold Abraham, Dr. Patrick Sebastian, Dr. J. Corbett McDonald, Dr. Arnold Brody, Dr. Morton Lippman, Dr. Nurtan Esmen, and Dr. Graham Gibbs. The list reads like a "who's who" in asbestos research. The implication, of course, is that these scientists' views should not go unheard.
2. Dissociation by the meeting chairman from the "Rhetorical excesses and procedural additives" of the meeting. Case was referring to public slander by audience members of scientists not present, which he said went uncommented upon by anyone in the chair. "Among the additives was the inviting of judges to portions of the meeting, the seeding of the audience with members of the White Lung Association, and the video taping of the conference by the labor group."[61]
3. Disclosure of conflict of interest for underwriters of the meeting. Case sent copies of this letter to the following people: Dr. Oakes Ames, President of the New York Academy of Sciences; Mr. Bill Boland, Director of Publications, N.Y.A.S.; Dr. Irving Selikoff, Honorary Life Governor, N.Y.A.S.; H. Kazemi, M.D., Massachusetts General Hospital; Dr. J.C. McDonald, London, U.K.; Dr. Brook Mossman, Burlington, VT; Dr. Graham Gibbs, Edmonton Alta; Dr. Jerome Abraham, Syracuse, N.Y.; and Dr. Jean Bignon, Cretail, France.

On November 20, 1990, Landrigan wrote an apologetic letter to J. Corbett McDonald. "I regret intensely the several unflattering remarks that were made about you and your work at the Collegium Ramazzini conference. Although clearly I had nothing to do with those remarks, I apologize for them in my capacity as Co-Chair of the conference."[62] Landrigan's list of people who were to receive this letter included: O. Ames, D.V. Bates, J. Begin, J. Bignon, M. Corn, J.M. Dement, E.A. Gaensler, J.B.L. Gee, F.D.K. Lidell, J. Merchant, B.T. Mossman, R.M. Rudd, A. Seaton, I.J. Selikoff, K. Tsuchiya, D. Wegman, H. Weil, and H.J. Woitowitz.

Brook Mossman wrote to Oakes Ames, Executive Director of the New York Academy of Sciences in December concerning the "Third Wave" conference. The whole letter is quoted here:

December 20, 1990

Oakes Ames, Ph.D.
Executive Director
The New York Academy of Sciences
2 East 63rd Street
New York, NY 10021

Dear Dr. Ames:

I was disappointed in your letter of November 27, 1990, specifically your refusal to extend an invitation to myself and the authors of the *Science* paper to contribute to the Proceedings and defend their views which were so misconstrued in the paper by Nicholson. It is now widely recognized in the scientific community that the Colloquium Ramezzini meeting on June 1, 990 was supported by the legal community with invitations extended by the organizers to judges presiding in major court cases. Through numerous communications and documentation, you and Mr. Boland are well aware of the purposeful exclusion of my co-authors and other scientists with different views than those expressed by Dr. Landrigan and colleagues at Mount Sinai. I doubt that inclusion of papers by Dr. McDonald and Gibbs will lend credibility to this meeting in view of these recent developments and Dr. Landrigan's deceitful letter to Dr. McDonald of November 20, 1990.

Dr. Case's letter of November 13, 1990 reinforces my previous correspondence on calls for peer review of the biased papers presented in June as well as disclosure of conflict of interest for underwriters of the meeting, specifically, individual and corporate legal entities who stand to benefit financially from the promulgation of certain points of view. I would appreciate (as would Dr. Case) a formal response from the Board as to whether the papers for the Proceedings will be refereed for this objectivity. Perhaps you can explain as well why the New York Academy of Sciences would publish the Proceedings of a biased meeting organized for other than scientific motives for which they have no knowledge of the attendees!

Sincerely,

Brook T. Mossman, Ph.D.[63]

Mossman sent copies of her letter to the following people: Dr. Landrigan, Mr. Boland, Dr. McDonald, Dr. Case, Dr. Bignon, Dr. Corn, Dr. Gee, Dr. Kazemi, Dr. Maltoni, Dr. Thomas, Dr. Sanders, Dr. Lichstein, Dr. Bruce Ames, and Dr. Abelson.

These are only a few of the angry letters of comments about the "Third Wave" conference. The rhetoric at the conference hardly made for open discussion or collegial feelings. Name calling and accusation of unethical behavior on both sides only pointed out the disarray and controversy within the scientific community over the issue of asbestos in buildings.

THE GREEN BOOK

Nevertheless, by the end of 1990 much had changed in the area of asbestos in public buildings. Indicative of the change is EPA new policy in this area. In July 1990, EPA published its fourth and last asbestos guidance book, *Managing Asbestos in Place: A Building Owner's Guide to Operations and Maintenance Programs for Asbestos-Containing Materials.*[64] Now the agency stressed operations and maintenance for buildings owners. The earlier guidance documents emphasized removal of asbestos. It was an about face change in policy. EPA rationalized the change by emphasizing what it called the Five Facts in *The Green Book* and in many other places. It is worth recalling the five facts to see how much the policy of *The Green Book* differs from the previous guidance documents. With publication of *The Green Book,* EPA quietly reversed its public policy position, one it had adhered to for over ten years. The Five Facts are presented in Chapter 8.

REFERENCES

1. USEPA, *Managing Asbestos In Place: A Building Owner's Guide to Operations and Maintenance Programs for Asbestos-Containing Materials,* OPTS, July, 1990.
2. HEI-AR, *Asbestos in Public and Commercial Buildings: A Literature Review and Synthesis of Current Knowledge,* HEI-AR, Cambridge, MA, 1991, 1.
3. Corn, M., Asbestos and disease: an industrial hygienist's perspective, *Am. Indust. Hygiene Assoc. J.,* 47, 522, 1986.
4. AIHA position statement on the removal of asbestos-containing materials (ACM) from buildings, *Am. Indust. Hygiene Assoc. J.,* 52, A-324, 1991.
5. Weill, H. and Hughes, J.M., Asbestos as a public health risk: disease and policy, *Ann. Rev. Publ. Health,* 7, 171, 1986.
6. Weill, H. and Hughes, J.M. Asbestos as a public health risk: disease and policy, *Ann. Rev. Publ. Health,* 7, 171, 1986.
7. *Report of the Royal Commission on Matters of Health and Safety Arising from the Use of Asbestos in Ontario*, Ontario Ministry of the Attorney General, Ontario, Toronto, 1984, 577.
8. *Report of the Royal Commission on Matters of Health and Safety Arising from the Use of Asbestos in Ontario*, Ontario Ministry of the Attorney General, Ontario, Toronto, 1984, 578.
9. Weill, H. and Hughes, J.M., Asbestos as a public health risk: disease and policy, *Ann. Rev. Publ. Health,* 7, 178, 1986.

10a. USEPA, Technical support document for regulatory action against friable asbestos-containing materials in school buildings, Preliminary Draft, Washington, D.C., 1980.

10b. USEPA, Support document for final rule on friable asbestos-containing materials in school buildings, Health effects and magnitude of exposure, Preliminary Draft, Washington, D.C., 1981.

10c. USEPA, Support document for final rule on friable asbestos-containing materials in school buildings, Health effects and magnitude of exposure, Preliminary Draft, Washington, D.C., 1982.

11. Weill, H. and Hughes, J.M., Asbestos as a public health risk: disease and policy, *Ann. Rev. Publ. Health,* 7, 190, 1986.
12. Weill, H. and Hughes, J.M., Asbestos as a public health risk: disease and policy, *Ann. Rev. Publ. Health,* 7, 181, 1986.
13. HEI-AR, *Asbestos in Public and Commercial Buildings: A Literature Review and Synthesis of Current Knowledge,* HEI-AR, Cambridge, MA, 1991, 2-1.
14. Letter from Archibald Cox, Chairman of the Board HEI-AR to Senator Barbara Mikulski and Congressman Robert Traxler, January 10, 1991.
15. HEI-AR, *Asbestos in Public and Commercial Buildings: A Literature Review and Synthesis of Current Knowledge,* HEI-AR, Cambridge, MA, 1991, 1.
16. HEI-AR, *Asbestos in Public and Commercial Buildings: A Literature Review and Synthesis of Current Knowledge,* HEI-AR, Cambridge, MA, 1991, i.
17. HEI-AR, *Asbestos in Public and Commercial Buildings: A Literature Review and Synthesis of Current Knowledge,* HEI-AR, Cambridge, MA, 1991, 1-1.
18. HEI-AR, *Asbestos in Public and Commercial Buildings: A Literature Review and Synthesis of Current Knowledge,* HEI-AR, Cambridge, MA, 1991, 1-1.
19. HEI-AR, *Asbestos in Public and Commercial Buildings: A Literature Review and Synthesis of Current Knowledge,* HEI-AR, Cambridge, MA, 1991, 1-1.
20. HEI-AR, *Asbestos in Public and Commercial Buildings: A Literature Review and Synthesis of Current Knowledge,* HEI-AR, Cambridge, MA, 1991, 8-10.
21. HEI-AR, *Asbestos in Public and Commercial Buildings: A Literature Review and Synthesis of Current Knowledge,* HEI-AR, Cambridge, MA, 1991, 1-10.
22. HEI-AR, *Asbestos in Public and Commercial Buildings: A Literature Review and Synthesis of Current Knowledge,* HEI-AR, Cambridge, MA, 1991, 1-9.
23. Proceedings of operations and maintenance programs in buildings containing asbestos, *Appl. Occupat. Environ. Hygiene*, 9(11) 1994.
24. Harvard University Energy and Environmental Policy Center, *Proceedings of Symposium on Health Aspects of Exposure to Asbestos in Buildings*, Harvard University. Cambridge, MA, 1989, iii.
25. Harvard University Energy and Environmental Policy Center, *Proceedings of Symposium on Health Aspects of Exposure to Asbestos in Buildings*, Harvard University. Cambridge, MA, 1989, 29.
26. Harvard University Energy and Environmental Policy Center, *Proceedings of Symposium on Health Aspects of Exposure to Asbestos in Buildings*, Harvard University. Cambridge, MA, 1989, 6.
27. Harvard University Energy and Environmental Policy Center, *Proceedings of Symposium on Health Aspects of Exposure to Asbestos in Buildings*, Harvard University. Cambridge, MA, 1989, 2.
28. Pacelle, M., Asbestos report funds cleanups unwarrented, *Wall Street Journal*, B1, Sept. 25, 1991.
29. Whysner, J., Covello, V.T., Kuschner, M., Rifkind, A.B., Rozman, K.K., and Trichopoulos, D., Commentary, asbestos in the air of public buildings: a public health risk?, *Prevent. Med.,* 23, 119, 1994.
30. Mossman, B.T., Bignon, J., Corn, M., Seaton, A., and Gee, J.B.L., Asbestos scientific developments and implications for public policy, *Science,* 247, 294, 1990.
31. Mossman, B.T., Bignon, J., Corn, M., Seaton, A., and Gee, J.B.L., Asbestos scientific developments and implications for public policy, *Science,* 241, 294, 1990.
32. Mossman, B.T., Bignon, J., Corn, M., Seaton, A., and Gee, J.B.L., Asbestos scientific developments and implications for public policy, *Science,* 241, 298, 1990.

33. Mossman, B.T., Bignon, J., Corn, M., Seaton, A., and Gee, J.B.L., Asbestos scientific developments and implications for public policy, *Science,* 241, 296, 1990.
34. Mossman, B.T., Bignon, J., Corn, M., Seaton, A., and Gee, J.B.L., Asbestos scientific developments and implications for public policy, *Science,* 241, 296, 1990.
35. Mossman, B.T., Bignon, J., Corn, M., Seaton, A., and Gee, J.B.L., Asbestos scientific developments and implications for public policy, *Science,* 241, 299, 1990.
36. Mossman, B.T., Bignon, J., Corn, M., Seaton, A., and Gee, J.B.L., Asbestos scientific developments and implications for public policy, *Science,* 241, 299, 1990.
37. Puncturing a panic, *Wall Street Journal*, Sept. 19, 1989.
38. *New York Times*, Jan. 19, 1990.
39. *Wall Street Journal*, Jan. 19, 1990.
40. *Time*, Jan. 29, 1990.
41. *Readers Digest*, Jan. 1, 1990 (condensed from *The American Spectator*).
42. *Forbes Magazine*, Jan.8, 1990.
43. *Forbes Magazine*, Jan. 8, 1990, 302.
44. USEPA, EPA Study of Asbestos-Containing Materials in Public Buildings: A Report to Congress, USEPA, Washington, D.C., February, 1988. p. 2.
45. *Forbes Magazine*, Jan. 8, 1990, 302.
46. *Forbes Magazine*, Jan. 8, 1990, 302.
47. Zitnansk, J.A., *Asbestos Abatement*, July/Aug. 1990, 9.
48. Zitnansk, J.A., *Asbestos Abatement*, July/Aug. 1990, 7.
49. *Chemical Regulation Reporter*, Bureau of National Affairs, Inc., Jan. 26, 1990.
50. Irving Selikoff, Press Release for Third Wave Conference, 1990.
51. Landrigan, P., in *The Third Wave of Asbestos Disease: Exposure to Asbestos in Place.* Annals of New York Academy of Science, Vol. 643, 1991, xv.
52. Landrigan, P., in *The Third Wave of Asbestos Disease: Exposure to Asbestos in Place.* Annals of New York Academy of Science, Vol. 643, 1991, xv.
53. Landrigan, P., in *The Third Wave of Asbestos Disease: Exposure to Asbestos in Place.* Annals of New York Academy of Science, Vol. 643, 1991, x.
54. *Congressional Record.* Vol. 136, No. 69. Washington, D.C., June 5, 1990.
55. Landrigan, P., in *The Third Wave of Asbestos Disease: Exposure to Asbestos in Place.* Annals of New York Academy of Science, Vol. 643, 1991, 228.
56. Landrigan, P., in *The Third Wave of Asbestos Disease: Exposure to Asbestos in Place.* Annals of New York Academy of Science, Vol. 643, 1991, 74.
57. Landrigan, P., in *The Third Wave of Asbestos Disease: Exposure to Asbestos in Place.* Annals of New York Academy of Science, Vol. 643, 1991, 100.
58. Letter from Bruce Case, Director of Center for Environmental Epidemiology, University of Pittsburgh Graduate School of Public Health. To Philip Landrigan, Editor of *The Third Wave of Asbestos Disease*, November 13, 1990.
59. Letter from Bruce Case, Director of Center for Environmental Epidemiology, University of Pittsburgh Graduate School of Public Health. To Philip Landrigan, Editor of *The Third Wave of Asbestos Disease*, November 13, 1990, 2.
60. Letter from Bruce Case, Director of Center for Environmental Epidemiology, University of Pittsburgh Graduate School of Public Health. To Philip Landrigan, Editor of *The Third Wave of Asbestos Disease*, November 13, 1990, 1.
61. Letter from Bruce Case, Director of Center for Environmental Epidemiology, University of Pittsburgh Graduate School of Public Health. To Philip Landrigan, Editor of *The Third Wave of Asbestos Disease*, November 13, 1990, 1.

62. Letter from Philip Landrigan, Editor of *The Third Wave of Asbestos Disease*. To J. Corbett McDonald, National Heart Lung Institute, London, Nov. 20, 1990. (Landrigan and copies of his letter to 18 people. The list reads like a who's who in asbestos research.)
63. Letter from Brook Mossman, University of Vermont to Oakes Ames, Executive Director, NYAS, Dec. 20, 1990.
64. USEPA, *Managing Asbestos In Place: A Building Owner's Guide to Operations and Maintenance Programs for Asbestos-Containing Materials,* OPTS, July, 1990.

8 Asbestos in School Buildings: Some Lessons

REVIEWING EVENTS

It took more than a decade to understand and widely disseminate knowledge about risks to health related to asbestos-containing materials (ACM) in buildings. During those years, non-occupational risks, although minimal, had been grossly exaggerated. Early decisions to remove asbestos were based solely on identification and the presence of asbestos and not on measurement of exposure to asbestos, or on information gathered following risk assessment procedures. Immediate removal, the least desirable option, resulted. It became the accepted policy articulated by the Environmental Protection Agency (EPA) in the first of its series of guidance documents.[1] *The Orange Book* reflected United States governmental concern about health risks for people who lived or worked in buildings containing asbestos in fireproofing, acoustic materials, lagging, and other products. EPA required schools to assess the potential for health risk from inhaling asbestos fibers by using an algorithm, later shown not to correlate with airborne levels in buildings.[2] The guidance documents led local communities to remove ACM from their schools The schools then sought funds and federal aid for removal. Former manufacturers of ACM were faced with lawsuits for removal costs, recovery of damages to building occupants, and punitive damages associated with failure to warn totaling billions of dollars. The asbestos removal industry and plaintiff's lawyers flourished. But EPA had initiated its guidance without environmental measurements of actual levels of asbestos in building air and without a risk assessment to place potential risks in perspective. They denigrated air sampling as a way to assess inhalation risk, and during the 1980s issued two more similar guidance documents, *The Blue Book* and *The Purple Book*.[3] All three guidance documents engendered concern, fear, and even hysteria.

At the same time, some members of the scientific and public health communities, unconvinced of the efficacy of EPA policy, began to investigate and report their own measurements of asbestos in the air of buildings. They utilized exposure data to perform risk assessments.[4] Nevertheless, the United States Congress passed the Asbestos Hazard Emergency Response Act in 1986 (AHERA) without even acknowledging that the information concerning risks to building occupants asbestos-in-air was exaggerated. EPA policy can be characterized as an instant solution to a complicated problem which ignored the need for science in determining environmental policy.

Reports of higher contamination in buildings after removal of ACM and measurements of asbestos-in-air in buildings with and without ACM accumulated along with measurements of asbestos content of outdoor air. The measurements led many scientists to question the wisdom of removal of ACM and slowed down the asbestos removal industry. The courts responded in a slower manner.

In 1990, the journal *Science* published a major scientific and policy challenge to the course of action for asbestos in schools which intensified the policy and scientific debate.[5] Discussions resulted about low levels of risk to building occupants and about options other than removal to deal with asbestos-in-buildings. The questions raised impacted on the asbestos removal industry, further reducing the number of removals. The 1990–1991 debate about ACM, including the HEI-AR Report and the Harvard Symposium, placed in abeyance a bill in Congress that would have extended AHERA to approximately 730,000 commercial buildings. Most importantly, EPA changed its stated policy in a new guidance document printed in 1990, *The Green Book.*[6]

THE GREEN BOOK

The Green Book, EPA's last guidance document for ACM in buildings, placed emphasis on managing ACM in place with an Operations and Maintenance Program (O&M) rather than removal. EPA articulated its policy change in the Five Facts paraphrased here.

- Fact one: Although asbestos is hazardous, human risk of asbestos disease depends on levels of exposure to airborne asbestos fibers.
- Fact two: Prevailing asbestos levels inside buildings are low, based on available data. Accordingly health risks to building occupants appears to be very low.
- Fact three: Removal of asbestos from buildings is often not the best course of action. In fact, an improper removal can create a dangerous situation where none existed previously.
- Fact four: EPA only requires asbestos removal during building renovation or demolition.
- Fact five: EPA does recommend in-place management (O&M) for ACM when discovered in buildings.[7]

The EPA policy stated in *The Green Book* did not ignore asbestos, it chose a program to preserve the integrity of asbestos instead of specifying removal as the only appropriate way to manage asbestos in buildings.

CONCLUSIONS

When the empirical base of science utilized for public health policy is in question, the manner in which the public and decision makers respond to an issue often become dependent on factors beside science. The presentation and characterization of public information then assumes great importance in how a risk is perceived and public health policy determined. In the case of non-occupational exposure to asbestos, the polarization of medical scientists and the regulatory climate of the 1980s paved the way for exaggerated public concern and misunderstanding of the issues. It led to a policy for asbestos in schools and other public buildings that was and still remains costly and controversial.

Issues

In retrospect, despite the complexity of the topics in this book, there is a central question which justifies the depth of concerns described about the initiation of government actions which caused hardship and disruption in schools, concern and fear among parents and teachers, and deep divisions within the scientific community. Does in-place ACM in school buildings cause increased risk to the health of school children and other occupants of school buildings? The government at first acted on the assumption that asbestos, a known occupational hazard, was also a hazard to occupants of buildings (non-occupational). Early decisions to act were based on this assumption and on the identification of ACM in a building. It resulted in wholesale removal without consideration of levels of exposure or determination of the extent of risk. Over time, the measurement of exposure and information about measurements of asbestos-in-air indicated airborne asbestos stirred up by removal persisted for long periods of time and created a greater risk in many buildings than if the asbestos had been left in place. In fact, removal workers faced greater risks to health than any occupants or workers in the building. One could speculate that exposure assessment caused EPA's shift in approach. More importantly, exposure assessment results confirmed the central concept of risk assessment. Risk is related to exposure.

The paradigm of occupational health and industrial hygiene professionals' recognition, evaluation, and control of a potentially hazardous material was not followed by EPA and others whose contrary approaches led to public hysteria. Much of the hysteria, confusion, wasted effort, and squandering of money could have been avoided if EPA had utilized the paradigm early in its policy determination. Instead they adjusted science to fit policy. The exposure assessments which began in the late 1970s subsequently reversed government policy. Furthermore, EPA's departure from the industrial hygiene, occupational health paradigm with its resulting public confusion, unnecessary abatement, staggering costs, and flood of litigation, left a permanent mark on the agency's scientific credibility. Surely these were unintended consequences.

Lessons

Public response to potentially toxic agents, such as asbestos, that were once demonstrated to have devastating effects in occupational settings, and present at lower concentrations of exposure in non-occupational settings, calls for greater public understanding of the low risk, for example, of asbestos inhalation to building occupants. The lesson is that people in key positions must understand the concepts of public health, especially the bedrock concept of toxicological science, "the dose makes the poison." The health experiences of workers occupationally exposed to extremely high concentrations of airborne asbestos in the past will not be duplicated when the public is exposed to much lower concentrations of the same toxic agent. A simple idea, but one not widely understood. The public needs a higher level of environmental understanding to have the capacity to evaluate environmental issues.

In the case of asbestos in school buildings, priorities were set by poorly informed public opinion, poorly informed public servants, and public perception, not on sound science. Sound science should be the underpinning of environmental policy in order to facilitate priority setting based on risk. William Reilly, EPA Administrator, articulated this issue in a speech to the American Enterprise Institute in 1990.[8] In the speech Reilly said that priorities were often set by partially informed public opinion, rather than the best judgment of relative risk.

> Without some way of determining relative levels of risk we would quickly become mired in a regulatory swamp. Legislators and regulators alike, not to mention the judicial system, can become absorbed in responding to public perceptions that are driven by the dramatic, the sensational and the well-publicized. And the price we pay for this responsiveness can be a diminished ability to deal effectively with less obvious, but perhaps more significant public health and environmental problems.[9]

The controversy over asbestos in buildings surely is an example of the clash between real risks and public perception. In the case of asbestos, critical policy decisions were political rather than scientific. Asbestos diseases are dose-related. There has been neither clinical or epidemiological support for asbestos-related disease from building occupancy.[10]

The Role of EPA

After publishing its last asbestos guidance document, EPA accepted part of the blame for the asbestos-in-schools fiasco. In a 179-page, in-house, document written in 1992 an internal task force reviewed the role EPA's communication policies and information played in asbestos policy decisions. The main EPA conclusion was that the agency's messages were misunderstood. EPA did not acknowledge the part they played in distorting the issue for the American public, and in ignoring the science that would have clarified the issue. This document did not and could not excuse the fact that critical environmental decisions which ignore science become essentially political ones as a direct result of EPA activities.

This was the EPA's conclusion. It is clear they attempted to worm their way out of responsibility for a miserably failed policy.

> Communicating about environmental risk is often a complex task. Communicating about hazards where there are divided opinions on the extent of risk and the effectiveness and costs associated with control make it even more difficult.
>
> Asbestos is a case in point. The hazards associated with asbestos, as with many environmental risks, come from exposure to the substance. If exposure is minimal, then the risk is minimal. When the substance is found almost everywhere — in

> thousands, if not millions of locations — then the evaluation of exposure becomes quite complex. The message is made more complex when the various alternatives proposed to minimize exposure — removal, enclosure, or encapsulation, management-in-place — are factored in. Finally, add into the equation the cost of control — the asbestos abatement industry is a $4 billion per year business. Who bears the burden of paying these costs — businesses, industry, consumers, tax payers?
>
> Such issues are at the heart of the asbestos problem, along with improved science, public relations campaigns by building owners and the asbestos industry, and lawsuits from parties seeking damages that may exceed $100 billion. It is this highly-charged atmosphere that EPA has had to communicate with a fearful public about asbestos.[11]

The writers of this document seem to have forgotten that EPA helped to create the "highly charged" atmosphere they complained about. EPA, a regulatory agency, chose and carried out environmental policy in conflict with the prevailing scientific consensus. They did not publicly defend the data on which they based policy and to this day they have not presented or discussed a risk assessment for asbestos in buildings. EPA has not publicly disclaimed the algorithm recommended in its guidance documents. Finally, the Office of Pesticides and Toxic Substances in EPA no longer contains an Asbestos Abatement Program. EPA is now out of the asbestos business and has been for some time. EPA declared victory, closed its asbestos abatement office, and left the mess it created to the states to clean up.

One would have thought that the first priority of EPA would have been to accurately assess the problem of asbestos. Failure to do so led to a costly policy. In the end, the cause of prudent environmental policy was ill served by action based on public perception driven by the dramatic and sensational rather than proven or sound scientific and technical tenets. The Congress seeking amendments to AHERA in 1990 finally got the message. They wanted to (1) facilitate public understanding of the relative risks associated with in-place ACM; (2) promote the least burdensome response actions; (3) describe the limited circumstances in which asbestos removal is necessary to protect human health; and (4) describe the relative risks associated with removal of asbestos-containing building materials.[12]

And what about the role of science in the agency? In 1990, Dr. Raymond Loehr, Chair of the Executive Committee of the EPA Science Advisory Board and Dr. Morton Lippmann, Chair of the EPA Indoor Air Quality and Total Human Exposure Committee, Science Advisory Board wrote on behalf of their respective committees expressing concern about their limited role in advising the EPA and Congress.

> At least one major indoor initiative (i.e., asbestos in schools) was undertaken by EPA without benefit of the advise of our Committee or the other advisory committees within the SAB. The scientific basis for the actions taken by EPA was not as thoroughly developed as those that have received review by your external scientific advisors. Substantial and legitimate scientific concerns that were subsequently expressed about the wisdom of EPA actions might well have been forestalled or ameliorated had such advice been sought. For example, recent papers in the *New England Journal of Medicine* (320:1721, 1989) and *Science* (247:294, 1990) examined the nature and magnitude of health risks from the inhalation of asbestos in relation to EPA's rules under AHERA.

> It was shown that the risk estimates were based on human experience with more toxic kinds of asbestos that account for less than 5% of the asbestos in place in U.S. buildings. Furthermore, the rules have led to removal decisions and procedures which have frequently increased rather than decreased human exposures. In a highly critical editorial in the March 2, 1990 issue of *Science*, Dr. Philip H. Abelson concluded that EPA's actions had reduced its credibility.[13]

The lesson is clear. A scientifically and technically insupportable position assumed by a government agency in its zeal to protect the public, as in the case of asbestos, could not hold up in the long run.

Once again, the nature of the debate led decision makers to depend on political factors rather than scientific ones in response to the issues. The presentation and the characterization of public information, we now know, assumed great importance in how risk was perceived and public health policy determined.

COULD IT HAPPEN AGAIN?

There are a host of other potentially toxic materials that have been demonstrated to cause adverse health effects at high exposure concentrations for those at work. Examples of these are lead, organic solvents, and mercury. The environmental exposures are orders of magnitude lower than those in the work environment. Not withstanding, the more susceptible nature of those in the general population when contrasted to healthy adult workers, EPA must focus on its risk assessments and the assumptions involved, and educate the public and the legislature before embarking on costly remedial programs that may also introduce additional risk to those conducting remediation. If they do not, the asbestos story will be repeated.

The nature of debate about environmental policy has been predominantly political, not scientific. There seems to have been precious little discussion about the nature of environmental protection, values, and ethics or about the relation between science and politics in an agency that still has not sorted out the proper role of each.

REFERENCES

1. USEPA, *Asbestos-containing materials in school buildings*, OTS, Washington, D.C., 1979.
2. USEPA, *Asbestos-containing materials in school buildings*, OTS, Washington, D.C., 1979.

3a. USEPA, *Guidance for controlling friable asbestos-containing materials in buildings*, OPTS, Washington, D.C., 1983.

3b. USEPA, *Guidance for controlling asbestos-containing materials in buildings*, OPTS, Washington, D.C., 1985.

4a. Corn, M., Crump, K., Farrar, D.B., Lee, R.J., and McFee, D.R., Airborne concentrations of asbestos in 71 school buildings, *Regul. Toxicol. Pharmacol.*, 13, 99, 1991.

4b. Crump, K. and Farrar, D.B., Statistical analysis of data on airborne asbestos levels collected in an EPA survey of public buildings, *Regul. Toxicol. Pharmacol.*, 10, 51, 1989.

4c. Hughes, J. and Weill, H., Asbestos exposure — quantitative assessment of risk, *Am. Rev. Resp. Dis.,* 35, 5, 1986.
5. Mossman, B.T., Bignon, J., Corn, M., Seaton, A., and Gee, B., Asbestos, scientific developments and implications for public policy, *Science*, 47, 294, 1990.
6. USEPA, *Managing Asbestos in Place,* P.T.S., Washington, D.C., 1990.
7. USEPA, *Managing Asbestos in Place,* Washington, D.C., 1990, V11–V111.
8. Reilly, W.K., Asbestos sound science and public perceptions: why we need a new approach to risk, American Enterprise Institute, Environmental Policy Conference. Washington, D.C., (Communications and Public Affairs, A107), June 2, 1990.
9. Reilly, W.K., Asbestos sound science and public perceptions: why we need a new approach to risk, American Enterprise Institute, Environmental Policy Conference. Washington, D.C., (Communications and Public Affairs, A107), June 2, 1990, 1.
10. Gaensler, E.A., Asbestos exposure in buildings, *Occupational Lung Dis.,* Clinics in Chest Medicine, 13, 231, 1992.
11. USEPA, *Communicating About Risk: EPA and Asbestos in Schools,* Final Report of the Internal Task Force, Jan. 1992.
12. *Hearing Before the Subcommittee on Toxic Substances, Environmental Oversight, Research and Development of the Committee on Environment and Public Workers,* United States Senate, 101 Congress, 2nd ed., S.1893, April 26, 1990.
13. Letter Sent to William Reilly, EPA Administrator from Morton Lippman and Raymond Loehr, Science Advisory Board, April 20, 1990.

Index

A

B

C

D

E

F

G

H

I

L

M

N

O

P

Q

R

S

T

U

V

W